Methods and Techniques for Cleaning-up Contaminated Sites

NATO Science for Peace and Security Series

This Series presents the results of scientific meetings supported under the NATO Programme: Science for Peace and Security (SPS).

The NATO SPS Programme supports meetings in the following Key Priority areas: (1) Defence Against Terrorism; (2) Countering other Threats to Security and (3) NATO, Partner and Mediterranean Dialogue Country Priorities. The types of meeting supported are generally "Advanced Study Institutes" and "Advanced Research Workshops". The NATO SPS Series collects together the results of these meetings. The meetings are co-organized by scientists from NATO countries and scientists from NATO's "Partner" or "Mediterranean Dialogue" countries. The observations and recommendations made at the meetings, as well as the contents of the volumes in the Series, reflect those of participants and contributors only; they should not necessarily be regarded as reflecting NATO views or policy.

Advanced Study Institutes (ASI) are high-level tutorial courses intended to convey the latest developments in a subject to an advanced-level audience

Advanced Research Workshops (ARW) are expert meetings where an intense but informal exchange of views at the frontiers of a subject aims at identifying directions for future action

Following a transformation of the programme in 2006 the Series has been re-named and re-organised. Recent volumes on topics not related to security, which result from meetings supported under the programme earlier, may be found in the NATO Science Series.

The Series is published by IOS Press, Amsterdam, and Springer, Dordrecht, in conjunction with the NATO Public Diplomacy Division.

Sub-Series

A.	Chemistry and Biology	Springer
B.	Physics and Biophysics	Springer
C.	Environmental Security	Springer
D.	Information and Communication Security	IOS Press
E.	Human and Societal Dynamics	IOS Press

http://www.nato.int/science
http://www.springer.com
http://www.iospress.nl

Series C: Environmental Security

Methods and Techniques for Cleaning-up Contaminated Sites

edited by

Michael D. Annable

University of Florida,
Gainesville, FL, U.S.A.

Maria Teodorescu

National Research and Design Institute for
Industrial Ecology - ECOIND,
Bucharest, Romania

Petr Hlavinek

Brno University of Technology,
Brno, Czech Republic

and

Ludo Diels

VITO NV, Mol,
Belgium

Published in cooperation with NATO Public Diplomacy Division

Proceedings of the NATO Advanced Research Workshop on
Methods and Techniques for Cleaning-up Contaminated Sites
Sinaia, Romania
9–11 October 2006

A C.I.P. Catalogue record for this book is available from the Library of Congress.

ISBN 978-1-4020-6874-4 (PB)
ISBN 978-1-4020-6873-7 (HB)
ISBN 978-1-4020-6875-1 (e-book)

Published by Springer,
P.O. Box 17, 3300 AA Dordrecht, The Netherlands.

www.springer.com

Printed on acid-free paper

CONTENTS

PREFACE

Soil and groundwater pollution caused by contaminants including petroleum hydrocarbons, represent a serious threat to the environment in NATO and NATO Partner countries. A major area of concern is oil production and large petrochemical sites. As an example, Romania has three major "hot spots" resulting from non aqueous phase liquid (NAPL) contamination situated in the north-eastern, southern and south-eastern geographical areas. These "hot spots" correspond to areas within which oil extraction and Romanian petrochemical industry operates (Prahova, Dolj, Bacau counties and the marine oil platform on the Black Sea). The largest contaminated area is in the southern part of Romania. There is historical pollution dating back approximately 60 years in this area. Since aquifer contamination is often a direct consequence of contaminated soils, it should be noted that the soils within the oil extraction areas are heavily impacted. This contamination has a direct impact on groundwater, but also affects flora, with further severe effects on the trophic chain (fauna, human population). Almost half of the accidental contaminant releases in Romania are caused by oil extraction or transportation (source: Romanian Waters – material for the Annual Report on the State of the Environment, 2005). These accidental releases have multiple causes, but most of them are related to the obsolete equipment, improper pipe conditions used for crude oil transportation and - in some cases - obsolete technological processes operating in the petrochemical industry.

As result of an environmental assessment for both the previous activities and the current operations, the Environmental Protection Agencies in Romania identified as priorities of major concern:

- in the short term – mitigation of the negative effects on the environment on a cost-effective basis, provided that the national environmental legal requirements are met;

- in the long term – to attain a level of performance regarding the environmental protection that approaches to the international requirements in the field.

Although these goals have been set, a realistic plan of implementation is not clear mainly because of the need to simultaneously address the environmental related issues and the limited capacity of the industrial operators to undertake the costs for both cleaning-up and more expensive technologies.

Using Romania here as an example NATO country, striving to join the EU in early 2007, these shortcomings need to be overcome. Thus a dedicated workshop was organized, incorporating the different expertise of researchers from NATO and NATO-Partner countries and including interested parties in the dialogue. The workshop was designed to support real progress in the area of contaminated site remediation. The results of the workshop were intended to be beneficial to neighboring countries in the region.

This publication comprises the presentations made at the NATO Advanced Research Workshop held in Sinaia, Romania 9 – 11 October, 2006. The contributions represent a wide range of issues and challenges related to contaminated site management from low cost solutions to petroleum contaminated sites to advances in biological treatment methods. This publication is meant to foster links between groups facing challenges cleaning up contaminated sites through presentations that explore the problems currently being addressed and solutions that are emerging in the field.

REMEDIATION OF METAL AND METALLOID CONTAMINATED GROUNDWATER

LUDO DIELS*, KAROLIEN VANBROEKHOVEN
VITO NV, Boeretang 200, 2400 Mol, Belgium

Abstract. This chapter describes the remediation of groundwater polluted by heavy metals. Special attention is paid to 'pump and treat' methods and to different *in situ* approaches. Emphasis is on microbial processes and their combination with physico-chemical systems.

Keywords: heavy metals; groundwater; water treatment; *in situ* bioremediation; *on site* treatment; soil washing and flushing; sulfate reduction; iron reduction; biosorption; bioprecipitation; immobilization; permeable reactive barriers; reactive zones; electron donors; sand filtration; bioreactors; upflow anaerobic sludge blanket reactor; natural attenuation

1. Introduction

In general more than 60% of contaminated sites in the world have problems with the presence of toxic metals such as cadmium, lead, copper, zinc, mercury and nickel. These metals are considered to be the most hazardous and are included on the US Environmental Protection Agency (EPA) list of priority pollutants (Cameron, 1992). Groundwater pollution is caused mostly by leaching of metals by infiltrating rainwater from the contaminated soil to the groundwater. Metal surface treatment activities have frequently caused pollution with toxic chromium (VI) and many large urban sites are contaminated by the same metal due to the leather industry (Saha and Ali, 2001). Other important industrial sectors that use heavy metals in their production processes are the non-ferrous industry and mining activities. In many cases, metals are solubilized by using acid process waters (e.g. sulfuric acid). This leads to groundwaters low in pH (between 2.5 and 6.5), rich in sulfate (between 100 and 5000 mg SO_4^{2-}/l) and high in dissolved metals (mostly in the range of 100 to 2000 μg/l). Other

* Ludo Diels, VITO NV, Boeretang 200, 2400 Mol, Belgium, email: ludo.diels@vito.be

M.D. Annable et al. (eds.), *Methods and Techniques for Cleaning-up Contaminated Sites*, 1–23.

co-pollutants may also be present (e.g. trichloroethylene used as degreaser in surface treatment). In the case of the mining industry, acid mine drainage (AMD) is responsible for widespread contamination. AMD is the result of a natural-occurring process when a metal sulfide mineral, particularly pyrite ores (e.g. chalcopyrite, sphalerite, etc.) are exposed to oxygen and water in the presence of naturally occurring sulfur-oxidizing bacteria such as *Acidothiobacillus ferrooxidans* and *A. thiooxidans* that act as biological catalysts (Johnson and Hallberg, 2005). The components of AMD have a deleterious influence on the biota of streams which receive it (Kontopoulos, 1988). In the case of uranium mines, uranium and radium are leached into the drainage water, poisoning the groundwater with radioactivity.

Toxic metals were distributed in the past via aerial-emissions (ancient pyrometallurgical processes) and have caused large-scale diffuse pollution. In this case a slow process of metal infiltration eventually will lead to leaching of metals into the surface water. As a first step the metals migrate from the unsaturated zone into the saturated zone, sorbing to aquifer material until all binding sites are saturated. Once the sorption capacity of the soil is exhausted, elevated metal concentrations can be transported through the groundwater and, in the long term, reach surface waters. Figure 1 illustrates metal distribution over different soil compartments as a function of time. Waste heaps and landfills can leak and lead to point sources leaching high concentrations of metals into the groundwater. Landfills containing jarosite, goethite, gypsum, slags or fines can, especially in the presence of organics, lead to solubilization of metals and metalloids, the contamination of surface water and contamination of groundwater due to anaerobic leaching. Metals from mines are more easily

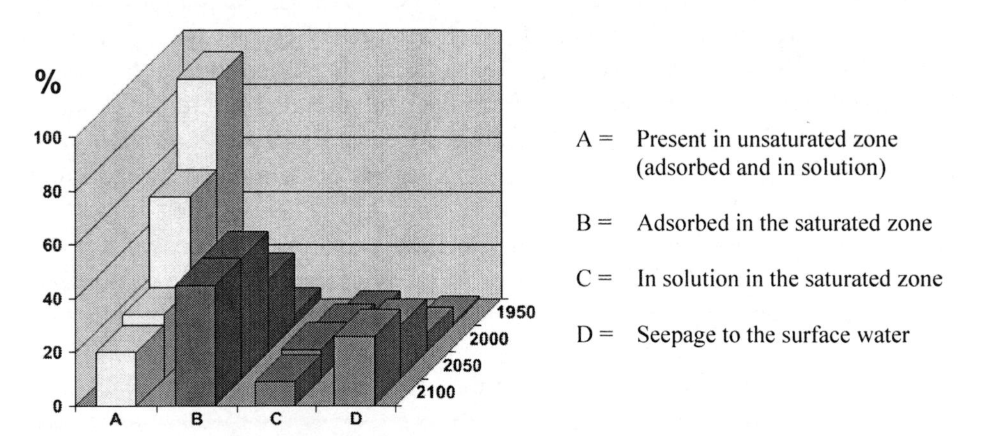

Figure 1. Expected evolution over time of the distribution of cadmium in the Kempen groundwater bodies if no remedial actions are undertaken (adopted from Schmidt, 2002)

transported by surface water and groundwater and rare events such as dam ruptures can lead to flooding of large areas contaminating the soil and the groundwater. An example of such a disaster was the mine tailings accident in Aznacollar (Grimault et al., 1999).

The current chapter reviews both full-scale and developing technologies that are available. The remediation of metal pollution can be based on extraction and physical separation, precipitation, immobilization and toxicity reduction. The selection of the most appropriate method depends on aquifer characteristics, pollutant concentration, types of pollutants to be removed, and the use of the contaminated medium. As an example, the evolution of the distribution of cadmium in groundwater bodies in The Kempen (Belgium) is presented in Figure 1. In order to avoid infiltration to the groundwater or seepage into the surface water, the metals can either be removed from the groundwater (decrease in C in Figure 1) or the metals can be adsorbed (immobilization) on the aquifer of the saturated zone (increase in B in Figure 1). In the approach, a distinction can be made between methods based on groundwater extraction, which remove metals from the water (decrease in C) or methods aimed at immobilizing the metals (increase in B) in the aquifer preventing further distribution of the contaminants. The first approach is an *on site* treatment method; the second approach an i method designed to reduce migration risks of the metals. Several physico-chemical and biological immobilization methods exist; here we will only focus on those methods that involve biological processes.

2. *On Site* Treatment Methods

The currently used pump and treat technology removes large volumes of groundwater from an aquifer and, if the water has to be returned to the aquifer, it must be treated in such a way that the metal concentrations fall below the standards for surface water or potable water. The above-ground treatment must lower the metal content to below the standards for surface water or drinking water. As these discharge standards are very stringent, a very expensive treatment will be necessary. If, after treatment, the water quality is inferior to the drinking water quality it has to be discharged into sewers or as surface water.

2.1. PUMP AND TREAT

Pump and treat methods are relatively easy to implement and control, but consume large pump energies, require high treatment costs and produce large amounts of waste products and water that must be discharged. A high volume discharge of water to the surface water can result in problems with ecotaxes. Also high-rate aquifer pumping can result in decreasing water tables, causing undesirable drainage of land or land subsidence.

Still, in some cases, pump and treat technology can be of interest for treatment of contaminant sources. An example is the full-scale groundwater treatment based on sulfate-reducing bacteria (Webb et al., 1998; Weijma J. et al., 2002; Greben et al., 2000). The system is composed of an UASB (Upflow Anaerobic Sludge Blanket) reactor with a three phase separator on the top. The excess of hydrogen sulfide that is produced is oxidized in a biological sulfide oxidation reactor. In this reactor, the bacteria and the sulfur adhere to the packing material until shear forces caused by the stirring action of the forced air stream detach the solids. Oxygen for the reaction is supplied as air and the carbon source (carbon dioxide) and nutrients (N, P) are present in the water. At high redox potentials and high oxygen levels, sulfide can be further oxidized to sulfate. To avoid an increased concentration of sulfate in the water, this reaction is minimized by controlling the supply of oxygen. A tilted-plate settler removes the solids that consist of sulfur produced in the biological oxidation process and metal sulfides and biomass flushed along with the effluent. A DynaSand filter further removes the suspended solids from the water. The method efficiently removes metals from the water, e.g. from 230 mg Zn/l to < 0.3 mg Zn/l. This THIOPAQ® system has proven to be reliable in treatment of sulfate containing, metal-contaminated water up to a scale of 400 m^3/h for more than 10 years.

Another biological treatment method is based on the use of microorganisms that induce biosorption and bioprecipitation on their surface. Biosorption is a biological treatment method which involves the adsorption of metals onto biomass such as algal, fungal or bacterial cells that can be dead or alive. If large-scale, inexpensive production techniques for the biomass are developed, this heavy metal treatment is promising.

Special biomass production can be avoided in the MERESAFIN (Metal Removal by Sand Filter Inoculation) process (Pümpel et al., 2001a; Diels et al., 2001). In this, bacteria able to biosorb or bioprecipitate heavy metals grow in a biofilm on a supporting material (e.g. sand). During contact with heavy metal-containing wastewater the biofilm adsorbs the metals. Subsequently the metal-loaded biomass is removed from the supporting material by the sand filter airlift and the resting biomass residual on the substratum can be re-used, after re-growth, for a subsequent treatment cycle. The supporting material can be sand or other materials retained within a moving bed sand filter which is based on a counter-flow principle (Figure 2). The water to be treated is admitted through the inlet distributor (1) in the lower section of the unit and is cleaned as it flows upward through the sand bed, prior to discharge through the filtrate outlet (2) at the top. The sand containing the heavy metals bound to the biofilm is conveyed from the tapered bottom section of the unit (3) by means of an airlift pump (4) to the sand washer (5) at the top. Cleaning of the sand starts in the pump itself, in which metal-loaded biofilms are separated from the sand grains by the turbulent mixing action. The contaminated sand spills from the pump outlet into the washer labyrinth (6), in which it is washed by a small flow of clean water. The metal-loaded flocs are discharged through the washwater outlet (7), while the grains of sand with a partly removed biofilm are returned to the sand bed (8). As a result, the bed is in constant downward motion through the unit. In this concept water purification and sand washing both take place continuously, enabling the filter to remain in service without interruption. In such a complete water treatment system groundwater is pumped through the Astrasand filter and purified. The wash water, containing the metal-loaded biomass, is drained to a lamella separator or settling tank. The water, coming from the thickener, is reintroduced in the sand filter. The sludge coming from the thickener is treated further in a filter press of lime. The filter cake (30% dry weight) obtained in this way, containing the metals (in some cases up to 10%), is recycled in a pyrometallurgical treatment facility (shaft furnace) of a non-ferrous company (Woebking and Diels, 2000).

1. Inlet distributor
2. Outlet
3. Dirty sand
4. Air-lift pump
5. Sand washer
6. Washer labyrinth
7. Wash water outlet
8. Cleaned sand

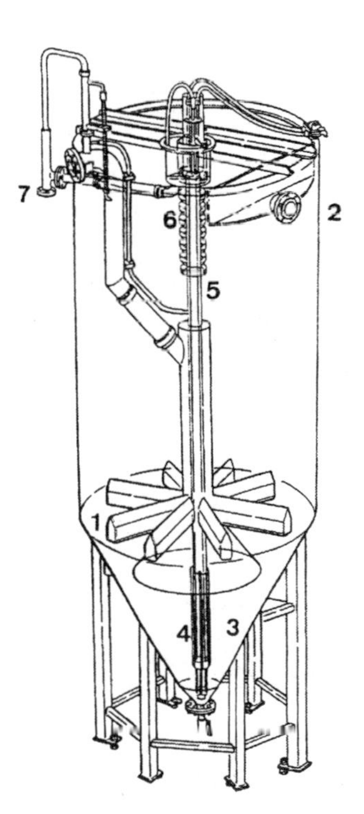

Figure 2. Moving bed sand filter concept (Diels et al., 2003)

Several other treatment technologies have been tested and prove OK at full scale or at pilot scale and described by Pümpel and Paknikar (2001). At he Homestake Mine at Lead, South Dakota, a rotating biological contactor (RBC) was immobilized with bacteria from the genus *Pseudomonas* that grow predominantly in biofilms. They are responsible for the degradation of free and metal complexed cyanide and thiocyanate and for the removal of heavy metals by biosorption in an aerobic process. Due to the slightly alkaline pH (7.5 to 8.5) and HCO_3^- produced within the biofilm there is a strong likelihood that precipitation of metal hydroxides and carbonates also contributes to metals removal following destruction of the metal-cyanide complexes.

The METEX® anaerobic sludge reactor is a cylindrical, Upflow Anaerobic Sludge Blanket (UASB) reactor filled with anaerobic sludge from standard sewage treatment plants. Slowly moving stirrers prevent the formation of short-circuit channels through the sludge bed, but keep the desired vertical gradients. From bottom to top subsequent zones with different metabolic activities and also different groups of microorganisms may develop in the sludge bed, depending on nutrients and on the electron acceptors available in the water (e.g. aerobic, denitrifying, sulfate reducing zone). The anoxic sulfate reducing zone is the most important one with respect to heavy metal removal in the METEX reactor, promoting the formation of highly insoluble metal sulfides. Further, bioprecipitation of metal carbonates, and biosorption/adsorption of dissolved metal species were shown to contribute to the overall metal removal process. The Bio-substrat® anaerobic micro-carrier reactor (Fürst and Burggräf, 2000) is similar to the METEX reactor but differs in two major aspects: the slowly stirred upflow reactor is filled with a granular microcarrier material with high sorption capacity (zeolite), and natural microorganisms, which have been adapted to the particular (ground)water matrix, are grown on the micro-carriers instead of using anaerobic sludge.

Wagner-Döbler et al. (2000) have demonstrated a bioreactor inoculated with a mixture of seven mercury-resistant, non-pathogenic *Pseudomonas* strains, isolated from mercury-rich environments. This reactor was especially designed for treatment of mercury-contaminated water.

2.2. SOIL WASHING AND FLUSHING

Metals, dissolved in the groundwater, are in equilibrium with the aquifer material which has a certain metal sorption capacity depending on the mineralogy and organic matter content. Since water solubility and desorption rates control metal removal from aquifers during pump and treat, additives are sometimes used to enhance water solubility and removal efficiencies. Metal desorption rates can be increased by a factor of more than 100-times by soil washing and flushing techniques. Soil washing and *in situ* flushing involve the injection of water with or without additives including organic and inorganic acids, sodium hydroxide (which can dissolve organic soil matter), water soluble solvents, e.g. methanol, non-toxic cations, and complexing agents, e.g. such as ethylenediaminotetraacetic acid (EDTA) and nitrilotriacetate (NTA). High clay and organic matter are particularly detrimental. Once the water is pumped from the soil, it must be extracted and then treated to remove the metals in wastewater treatment facilities or re-used in the flushing process (Mulligan et al., 2001). In general, soils with low contents of cyanide, fluoride and sulfide, a CEC of 50-100 meq/kg and particle sizes of 0.25-2 mm, with contaminant water solubility larger than 1000 mg/l, can be most effectively cleaned by soil washing (Mulligan et al., 2001).

Several groundwater treatment technologies exist such as sodium hydroxide or sodium sulfide precipitation, ion exchange, activated carbon adsorption, ultrafiltration, reverse osmosis, electrolysis/electrodialysis and biological systems (Patterson, 1985). As mentioned earlier, biosorption and bioprecipitation methods can be used. Merten et al. (2004) used *Escherichia coli* and fungal (*Schizophyllum commune*) biomass to adsorb uranium and rare earth elements from seepage water from a former uranium mining site in Eastern Thuringia in Germany.

A special approach is necessary when organic complexing agents (e.g. EDTA, NTA, sophorolipids) are used. The treatment method can be based on bacterial breakdown of the organic component followed by adsorption of the metal to the biomass in the water treatment plant. The problem is that very often these complexes are stabilized by the metal and biodegradation, normally feasible with the sodium or magnesium complex, will be prevented by the metal. However, in some cases bacteria can be isolated that are able to cleave heavy metal-organic complexes (Francis and Dodge, 1993).

3. *In Situ* Treatment Methods

Heavy metals can occur in several forms in the groundwater and on the aquifer. In many cases the aquifer-groundwater zone is stratified, having layers that are aerobic, nitrate-reducing, iron-reducing and/or sulfate-reducing depending *on site*-dependent circumstances. *In situ* treatment aims at immobilizing the metals in the aquifer. The methods used to cause precipitation determine the exact reaction mechanisms occurring, which in their turn determine the long-term stability of the metal precipitates. The technique is only acceptable as a viable remedial option if very stable precipitates are formed.

To determine the speciation of metals in aquifers, specific extractants are used since they are supposed to solubilize metal fractions present in different chemical environments (each with a specific availability/solubility). By sequentially extracting the aquifer with solutions of increasing strength, a more precise evaluation of the different fractions can be obtained (Tessier et al., 1979). An aquifer is shaken over time with a weak extractant, centrifuged, and the supernatant is removed by decantation. The pellet is washed in water and the supernatant removed and combined with the previous supernatant. Next the procedure is repeated with a stronger extractant. Extraction reagents can be (from weak to strong): water, $MgCl_2$, sodium acetate, hydroxylamine and ammonium acetate, HCl. This procedure allows determination of the leachable/ exchangeable fractions, carbonate fraction, Fe-Mn-oxide (reducible) fraction and organic fractions (oxidizable fraction).

3.1. NATURAL ATTENUATION

Natural attenuation is a process in which metals are immobilized by naturally-occurring chemical, biological and physical processes. Metals can be complexed by binding to carboxylic or phenolic groups of humic acids (Fe=Cu>>Zn>>Mn) and can precipitate as hydroxides, oxides, carbonates, phosphates and sulfides. Microbial sulfate reduction can lead to the precipitation of metal sulfides: some metalloids such as As can co-precipitate in Fe_2O_3 or MnO_2. Other metals can adsorb to $Fe(OH)_3$. Metals can also be taken up by plants. Suspended or colloidal materials can be filtered by their passage through the soil matrix while alkalinity generation by, e.g. dolomite or calcareous materials, can lead to pH increase and subsequent metal precipitation. Microorganisms can also play an important role in adsorbing or bioprecipitating metals and also influence the toxicity and speciation. Microorganisms can oxidize some metals such as iron and manganese and make them insoluble. Arsenic can be oxidized from arsenic (III) to arsenic (V) making it less toxic. Arsenic (V) will also co-precipitate on/with iron. Chromium (VI) can be reduced to chromium (III) which is more insoluble and less toxic.

The consulting company Tauw developed (with the support of SKB in the Netherlands) an expert system called BOSS for evaluation of natural immobilization of heavy metals in aquifers. This system requires the following parameter inputs: $Fe^{2/3+}$, SO_4^{2-}, HCO_3^-, DOC, soil organic matter, clay, and iron and aluminum oxides (Steketee, 2004).

3.2. *IN SITU* BIOPRECIPITATION

In situ metal (bio)precipitation (ISMP) is the process in which sulfate-reducing bacteria are grown by the addition of electron donors as molasses, lactate, HRC® (Koenigsberg et al., 2002), MRC® (Koenigsberg, 2002), ethanol and/or other carbon sources. The bacteria oxidize the electron donor and use the released electrons to reduce the sulfates present in the water. The formed sulfides cause a precipitation of the metals from solution (Hao, 2000; Janssen and Temminghoff, 2004).

A few conditions are required for the process. Sulfate reducing bacteria must be present in the aquifer. Sulfate must be present in sufficient concentration (mostly > 200 mg /l). A not too extreme pH (5 – 8) is necessary as is a minimum content of nutrients (N and P), no oxygen and a low redox potential. The general principle is presented in Figure 3. The following reactions can take place:

$$SO_4^{2-} + 8\ e^- + 8\ H^+ ==> S^{2-} + 4\ H_2O$$
$$CH_3COOH + 2H_2O ==> 2CO_2 + 8\ H^+ + 8\ e^-$$
$$CH_3COOH + SO_4^{2-} ==> 2HCO_3^- + HS^- + H^+$$

$$H_2S + Me^{++} ==> \underline{MeS} + 2H^+$$

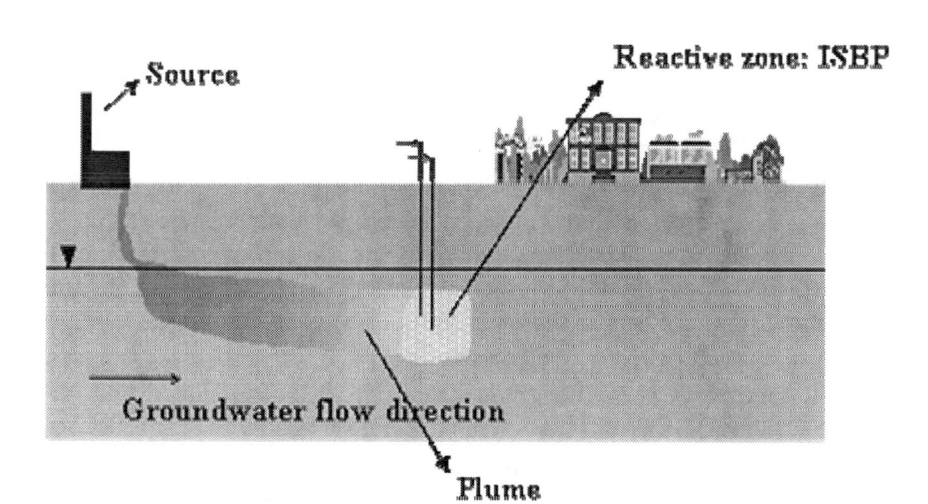

Figure 3. General principle of an *in situ* bioprecipitation (ISBP) treatment method

Several lab-scale tests (batch and column tests) are currently available to study the feasibility of the process. However, only a few field tests have been performed up to now. One field test was at a non-ferrous industrial site in Belgium, with groundwater contamination of Cd, Zn, Ni and Co. Another site in Belgium, contaminated with chromium (VI), was treated by molasses injection in order to reduce chromium (VI) to chromium (III). A third demonstration was obtained at a metal surface treatment site, contaminated by Zn, in Dieren in The Netherlands.

The following part will give a summary of results and important parameters (Diels et al., 2005a). The sulfate concentration is important in relation to the electron donor. If the sulfate concentration is low (< 200 mg SO_4^{2-}/l) the sulfate-reducing process will not start unless a sulfate-reducing inoculum is added. Further, at low sulfate concentrations the sulfate reduction could only be started if hydrogen was added as electron donor. The electron donor concentration is important since an excess can lead to methanogenic conditions. In addition, high concentrations of acetate could lead to inhibition of certain SRB strains. Moreover, the use of acetate always leads to a very slow precipitation process because only few SRBs have the ability to use acetate as carbon source. High concentrations of molasses can cause a pH decrease due to fermentation processes. A lowering of pH can lead to metal release from the aquifer into the groundwater. High molasses concentrations also lead to the complexation of chromium (III) and hence to its solubilization. A wide range of electron donors has been proved useful in the process, varying from expensive pure substrates such as ethanol, lactate (Hammack and Edenborn, 1992), and hydrogen (Van Houten et al., 1994) to economically more favorable waste products, with or without enrichment with pure substrates or inoculation with monocultures or media (manure, sludge, soil) containing SRB (Maree and Strydom, 1987; Prasad et al., 1999; Annachhatre and Suktrakoolvait, 2001).

Sulfate-reducing bacteria can be enriched at a pH between 4 and 8. At neutral pH, SRB from different origins could be detected (Groudev et al., 2005a). At low pH only *Desulphotomaculum* and *Desulphorosinus sp.* could be detected (Geets et al., 2004). Molecular biology techniques can be used to identify the SRB and study their diversity. Either 16 S rRNA gene-based primers can be used or *dsr*-based (dissimilatory sulfite reductase) primers, one of the prime enzymes in sulfate reduction. It has been concluded that only by using *dsr*-based DGGE a real biodiversity could be observed. Temminghof and Janssen (2005) stated that at a location in The Netherlands (Dieren) with an initial pH of 3.9, the ISBP process could only be started after addition of an SRB inoculum. A redox potential < -200 mV is necessary to grow the SRBs. Temminghof and Janssen (2005) used Na_2S to reduce the E_h. Diels et al. (2005b) used other redox manipulating compounds to reduce the E_h, especially in low pH conditions.

Janssen and Temminghoff (2004) discuss the need for specific SRB medium to stimulate the growth of SRB. This indicates that, in specific cases, nutrients (N and P) need to be added and sometimes also other trace elements. From our experience, we can say that at appropriate redox potential and pH, the sulfate reducing process can be induced in nearly all aquifers.

The ISBP process was investigated for Cu, Zn, Cd, Ni, Co, Fe, Cr, and As. The first field tests showed that ISBP is feasible as a strategy for sustaining groundwater quality (Ghyoot et al., 2004). However, some questions remain to be answered, especially about the pH decrease due to molasses fermentation, the stability of the Ni and Co precipitates and the type, amount and injection frequency of electron donor.

In order to define the stability of the immobilized metals it is important to analyze the metal precipitates by sequential extraction (Tessier et al., 1979). A detailed study was produced by Diels et al. (2005b). A summary of these results is given below. An aquifer from a non-ferrous contaminated site was used in a column study. Groundwater from the site was pumped over the aquifer-loaded columns over two years. The columns were treated with different electron donors including molasses and lactate. Besides lactate, lactate containing additional nutrients N and P (a mixture of ammonium nitrate and ortho-phosphate) was also added in order to avoid nutrient limitation for the SRB population. A control column without electron donor was also operated which was representative of a natural attenuation process (NA). In the electron donor amended columns, the redox potential decreased to –250 mV, the initial pH of about 4 went up to 6 and most of the metals were removed. In the NA column metals were not removed. Different carbon sources like molasses, HRC® and lactate (with or without N/P nutrients) promoted sulfate reduction within 8 weeks, with average sulfate-removal efficiencies of 50%. At the same time, substantial removal of Cd, Zn (at least 75%) and Co (at least 45%) took place, whereas attenuation of Ni was still difficult and unstable. In the following weeks, molasses failed to maintain sulfate reduction, and metal concentrations quickly increased in the column effluent. After 15 weeks, the sulfate reduction process in the HRC® amended column also showed a decreasing efficiency of metal removal, whereas the ISBP process seemed to be stable in the lactate-amended columns (Vanbroekhoven et al., 2005a). Table 1 presents the results of the sequential extraction for Zn of the column operated with molasses as substrate and compares the results of sequential extraction of Zn of the aquifer at T0. At T2, after two years of operation, the columns were stopped and samples were taken at four places (C1, C2, C3 and C4) in the column. C1 is the inlet of the column and C4 is the outlet, and C2 – C3 intermediate. Different extractions were made in order to define the speciation of the metal: leachable fraction (extraction with water), exchangeable fraction (extraction with $MgCl_2$), carbonate fraction (extraction with sodium acetate), Fe-Mn oxide fraction or reducible fraction (extraction with $NH_4OH.HCl$), organic or oxidizable fraction (extraction with NH_4-acetate) and the residual fraction. The sum of all the fractions was compared with a second sample treated by *aqua regia* (HCL and HNO_3) in order to control the mass balance. It was found that the mass balance was good since the aquifer was not homogeneously contaminated. There was a large difference (increase of metal deposition) between T0 and T2 as during those two years metals were precipitated continuously on the aquifer. The metal concentrations were highest at the inlet (C1) of the column as there the oxygen was first consumed and redox potential decreased first. Leachable, exchangeable and carbonate fractions were decreased compared to the original situation. This indicated that all metals had moved into a more tightly-bound precipitate. Nearly all the metals were found in the reducible (Fe-Mn oxide) and mostly in the oxidizable fraction (organic). In fact, the metal sulfide precipitates were expected in the oxidizable fraction.

TABLE 1. Sequential extraction of Zn (mg/kg dm) from aquifer derived from the column operated with molasses as electron donor

Sample	Leachable	Exchangeable	Carbonate	Fe-Mn oxide	Organic	Residual	Total Seq. Ext.	Total Extraction
T0	227	809	93	144	69	75	1417	1630
C1	71	262	51	689	3 153	356	8583	8400
C2	80	308	38	625	2 783	287	4121	4520
C3	77	281	41	571	1 536	201	2708	3410
C4	42	155	20	422	2 647	260	3546	3950

T0 is the metal distribution over the different fractions a the start up of the column experiment at time zero. C1, C2, C3 and C4 present the sequential extractions of Zn in the columns under respectively abiotic conditions, natural attenuation, with molasses and fed with lactate (+N/P).

Table 2 presents a comparison of sequential extractions for four columns operated under abiotic conditions, natural attenuation conditions or with molasses or lactate (+N/P) as the electron donor. The leachable and exchangeable fractions were stable in the NA column and were reduced in the molasses and lactate columns. However the reduction was always higher in the lactate column compared to the molasses column. The same was true for the carbonate fractions, but the differences were smaller. It was also observed that the Zn was precipitated in the reducible fraction and the highest amount was recovered from the oxidizable fraction (ZnS). However, for Ni, this was recovered in the reducible and mostly in the oxidizable fraction in the lactate-amended column. Only very small amounts of Ni could be recovered from the aquifer of the molasses-amended column. Here a large difference was observed between molasses and lactate as electron donors. Lactate always tends to precipitate the metals in a more stable form and this was especially true for Ni. Ni removal and stable precipitation is rather poor in the case of molasses-amended columns while optimal conditions could be obtained with lactate. A complete discussion of these results is presented in Diels et al. (2005b).

TABLE 2. Sequential extraction for Zn and Ni from aquifer-derived material from columns operated with molasses or lactate as electron donor compared to natural attenuation

	Zn (mg/kg dm)			Ni (mg/kg dm)		
	NA	Molasses	Lactate N/P	NA	Molasses	Lactate N/P
Leachable						
T0	227	227	227	81	81	81
C1	267	71	16	93	77	103
C2	306	80	1	111	75	35
C3	292	77	1	108	75	12
C4	293	42	2	107	76	10
Exchangeable						
T0	809	809	809	22	22	22
C1	763	262	45	23	23	81
C2	890	308	0	31	37	46
C3	828	281	0	62	46	19
C4	720	155	0	21	39	17
Carbonate						
T0	93	93	93	22	22	22
C1	119	51	83	23	23	81
C2	124	38	9	31	37	46
C3	191	41	6	62	46	19
C4	75	20	2	21	39	17
Fe-Mn oxide						
T0	144	144	144	5	5	5
C1	189	689	445	189	11	126
C2	199	625	430	7	11	95
C3	198	572	381	7	12	39
C4	265	423	228	26	16	32
Organic						
T0	69	69	69	3	3	3
C1	69	3153	2652	4	6	3425
C2	83	2783	1594	5	7	1706
C3	91	1536	684	5	10	794
C4	99	2647	1217	8	44	530

In situ precipitation of metals from contaminated groundwater by acceleration of biogeochemical processes that may occur naturally, is a promising sustainable technology to remediate sites polluted by metals. However, often these aquifers contain high concentrations of Fe, presumably present as Fe(III) minerals which may compete with the *in situ* bioprecipitation process by sulfate-reducing bacteria. Based on thermodynamics, microbes are supposed to use the electron acceptors resulting in the highest energy yield. Once oxygen is completely used or becomes limiting, microbes utilize nitrate, followed by iron and later on sulfate. Dissimilatory iron reduction has indeed often been assumed to be a competitive process in our studies since high concentrations of Fe-likely to be Fe(II) at neutral or slightly acidic pH have been measured in the groundwater used in ISBP process. Often it could be found that the iron reduction took place in the beginning of the stimulation of the process at redox potentials higher than -250 mV. In fact by using specific 16S rRNA gene fragments as primers, Geobacteriaceae could be detected (Vanbroekhoven et al., 2006). In the *in situ* processes the alkalinization, as a result of the iron or nitrate reduction process, aided in the precipitation of the metals as hydroxides or carbonates and explains why not always the metals are found in the oxidizable fraction but sometimes in the reducible fraction.

3.3. PERMEABLE REACTIVE BARRIERS

Permeable Reactive Barriers (PRB) are installed perpendicular to the groundwater flow direction. The barrier is composed of a reactive material that allows the removal of the pollutant (e.g. heavy metals) from the groundwater. The PRB filler material can be an adsorbent (e.g. silicates, zeolites, hydroxyapatite) to which the metals can bind via ion exchange to functional groups or precipitate via ligand complexation processes (Benner et al., 1997; Waybrant et al., 2002). The material can also be limestone that leads to precipitation of metals due to pH-neutralization. The PRB filler material can also be zero-valent iron. This material can remove metals by the processes of reduction, sorption or cementation. In reduction, metals can be reduced by the electrons that escape due to oxidation of the iron. This is the case for the reduction of soluble Cr (VI) to the less soluble and less toxic Cr (III) (Powell et al., 1995). Uranyl ions [U(VI)] can be reduced to non-soluble uraninite [U(IV)]. Metals can also just sorb onto the iron surface. Metal ions like Cu^{2+}, Ni^{2+} and Cd^{2+} can be reduced to their zero-valent form by oxidation of the iron at the same time and allowing the electrons to travel from the iron to the metal. This process is known as ‘cementation’. Diels et al. (2005a) compared different materials to induce cementation. An organic material such as peat was compared with an iron oxide (Ferrosorp), zero-valent iron (ZVI), hedulite (a titanium oxide waste product) and a commercial arsenic-sorbent ('P.As'). The analysis was carried out after 6 hours incubation. Metals could be removed from the leachate by Ferrosorp, ZVI and 'P.As'. In the case of hedulite it was observed that some metal (Zn) was leached into the water. Other newer results were presented by Van Roy et al. (2005a). Munro et al. (2004) described a PRB filled with Bauxsol™ mixed with sand (in order to keep a high permeability). Bauxsol™, a product made from seawater-neutralized red mud (a by-product of aluminia refining) buffers the pH at 8.8 and has been shown to remove > 99% of heavy metal loadings > 1000 meq/kg, which would make it an ideal medium for PRB. Bastiaens et al. (2005) proposed an alkalinity generating PRB, based on crushed limestone, which can help to increase the pH in order to start biological reactions.

On the other hand, PRB can be filled with organic materials or combinations of materials in order to allow bacterial processes to take place. These processes involve sulfate and iron reduction, oxidation and reduction processes and adsorption processes. Groudev et al. (2004, 2005a) studied the removal of uranium and other heavy metals from an acid mine drainage contaminated groundwater in a so-called "Multibarrier" system. Gilbert et al. (2002) have described the combination of several materials in a PRB for treatment of a metal contaminated groundwater.

3.4. COMBINED CHEMICAL AND MICROBIAL PROCESSES

Acid mine drainage (AMD) waters contain a mixture of contaminants. For example this is the case at the Curilo deposit near Sofia in Bulgaria. The AMD contains radionuclides (uranium, radium), toxic heavy metals (mainly iron, manganese, copper, zinc and cadmium) and sulfates. Groudev et al. (2005a) have reported the use of a so-called 'Multibarrier' PRB in which different passive water treatment processes are combined either successively (i.e. sequentially) either mixed. The first system described was a successive combination of four units. The first unit (I) was an aerobic barrier in which most of the ferrous iron was turned into the ferric state as a result of oxidations carried out by acidophilic chemolithotrophic bacteria (*Acidothiobacillus ferrooxidans* and *Leptospirillum ferrooxidans*). The second unit (II) was a barrier in which most of the ferric ions were precipitated as $Fe(OH)_3$ as a result of chemical neutralization in the presence of crushed limestone. The third unit (III) was an anaerobic barrier for microbial dissimilatory sulfate reduction. It contained a mixture of slowly biodegradable solid organic substrates (plant and spent mushroom compost, cow manure, sewage sludge, hay) and was inhabited by a consortium of sulfate-reducing bacteria and other metabolically interdependent-microorganisms. In this barrier, the non-ferrous metals were precipitated mainly as insoluble sulfides, and uranium was precipitated as uraninite (UO_2) as a result of the reduction of the hexavalent uranium to the tetravalent form. Radium was removed mainly as a result of adsorption by the organic matter and clay minerals present in the barrier. The effluents from this barrier were enriched in dissolved organic compounds but still contained manganese in concentrations higher than the permissive levels. In the last unit (IV), the dissolved bivalent manganese was oxidized under aerobic conditions by different heterotrophic bacteria to Mn^{4+}, which was precipitated as MnO_2. The dissolved organic compounds were also removed by the heterotrophs inhabiting this barrier. Table 3 presents the results of the water after treatment in the permeable Multibarrier. Groudev et al. (2005b) isolated the predominant organisms occurring in the different units: Fe^{2+}-oxidizing chemolithotrophs in the first unit: nearly no bacteria were found in unit II, cellulose-degrading, sulfate reducing

TABLE 3. Metal concentrations and physico-chemical parameters of drainage water during their treatment in a Multibarrier at the Curilo deposit

Parameters	Before treatment	Effluents from the barriers			
		I	II	III	IV
pH	2.71 - 2.90	2.73 - 3.41	4.55 - 5.10	7.25 - 7.58	7.32 - 7.65
Eh, mV	(+325) - (+484)	(+488) - (+594)	(+240) - (+293)	(-235) - (-260)	(+257) - (+286)
Dissolved O_2, mg/l	2.4 - 2.8	4.6 - 5.3	2.1 - 2.5	0.2 - 0.4	2.4 - 3.0
TDS, mg/l	1184 - 1720	1144 - 1680	820 - 1076	532 - 684	541 - 701
Solids, mg/l	28 - 59	27 - 64	46 - 104	35 - 77	37 - 71
Dissolved organic C, mg/l	0.6 - 0.9	0.7 - 1.0	0.6 - 0.8	51 - 140	14 - 21
Sulfates, mg/l	671 - 932	664 - 923	532 - 695	293 - 406	302 - 410
U, mg/l	2.40 - 3.87	2.30 - 3.61	1.64 - 2.75	<0.1	<0.1
Ra, Bq/l	0.35 - 0.55	0.35 - 0.50	0.25 - 0.35	<0.05	<0.05
Cu, mg/l	1.40 - 2.84	1.37 - 280	1.04 - 2.13	<0.1	<0.1
Zn, mg/l	12.5 - 20.8	12.2 - 19.9	11.6 - 17.0	<1.0	<1.0
Cd, mg/l	0.06 - 0.10	0.06 - 0.10	0.04 - 0.08	<0.01	<0.01
Pb, mg/l	0.28 - 0.64	0.28 - 0.60	0.25 - 0.53	<0.05	<0.05
Ni, mg/l	0.91 - 1.78	0.90 - 1.7	0.82 - 1.52	<0.1	<0.1
Co, mg/l	0.71 - 1.50	0.70 - 1.43	0.64 - 1.22	<0.1	<0.1
Fe, mg/l	257 - 590	251 - 578	21 - 41	<1.0	<1.0
Mn, mg/l	8.2 - 20.3	8.1 - 19.8	7.9 - 17.6	0.5 - 3.5	0.2 - 0.8
As, mg/l	0.23 - 0.45	0.21 - 0.41	0.14 - 0.28	<0.05	<0.05

and methanogenic bacteria occurred in unit III, and aerobic heterotrophic bacteria (with some $S_2O_3^{2-}$-oxidizing chemolithotrophic bacteria) were found in unit IV.

In another approach Groudev et al. (2005b) combined an alkaline limestone drain with an anaerobic section. The effluent from the barrier was allowed to flow through a wetland system used as a polishing step before final discharge of the mine drainage water to surface water. The wetland removed small traces of metals by adsorption and uptake by plant material. However, plant uptake was very important in removing radium (between 110 and 610 Bequerel/kg dry plant matter). Uranium can be reduced by SRBs and also some combinations with zero-valent iron in which the iron reduces the uranium to an insoluble form of uraninite, can be mentioned as interesting new emerging technologies (Mallants et al., 2002).

The Bitterfeld site in Germany is a large industrial site contaminated by chlorinated aliphatics and aromatics. At some locations heavy metal pollution is also present. It is well known that chlorinated aliphatics can be dehalogenated under sulfate-reducing or methanogenic conditions (El Fantroussi et al., 1998). However in the simultaneous presence of CAH and heavy metals it was shown that no degradation took place, presumably due to toxicity of the metals toward the dehalogenating bacteria. Vanbroekhoven et al. (2005b) found that no detectable amounts of PCE degrading bacteria were present. The addition of a dehalogenating inoculum led to dehalogenation only in the absence of metals. Only after addition of a suitable medium to induce sulfate-reducing bacteria could CAH dehalogenation and metal sulfide precipitation take place.

A groundwater, contaminated by a landfill in Poland (Tarnowkie Gory), contained zinc, copper and sulfate, and also boron, barium and strontium. The quaternary aquifer was composed either of sand or clay. In both aquifer textures, the ISBP could be used to immobilize the metals in the presence of molasses, lactate or HRC®. However B, Ba and Sr were not removed and even lead to release in the presence of molasses (pH decrease) on the sandy aquifers (Diels et al., 2005a; Van Roy et al., 2005b). Therefore, some adsorbents were added to the test systems, yielding a combined system of adsorption and ISBP. The results are presented in Table 4. Here Sr, and Ba were removed completely due to adsorption on zeolite and B to about 35%. In case of boron, a combination of different adsorbents was necessary to remove the different chemical forms of boron (e.g. borate). Zeolite was successful to some extent and the combination of anionic Metasorb with Apeyron PAsXP lead to relatively good B adsorption. Dolomite as an adsorbent had no effect on the removal process. It was also observed that the ISBP was slightly inhibited by the addition of zeolite. In these tests also MRC® (Metal Removing Compound, Regenesis) was used and similar results were obtained as with HRC®.

TABLE 4. ISBP of non-ferrous metals and B, Ba and Sr in mesocosms containing groundwater and aquifer in combination with adsorbents

Metal	No addition	HRC®, MRC®, lactate, molasses	C-source + dolomite, diatomaeous earth	C-source + zeolite
SO_4	-	+++	+++	+
Zn	-	+++	+++	+
Sr	-	-	-	+++
B	-	-	-	0.35
Ba	-	-	-	+++

+++: very fast; ++: fast; ± removal; - no removal

C-source = molasses or HRC®

4. Concluding Remarks

This chapter attempts to give an overview of the different existing or developing methods for treatment of groundwater contaminated with heavy metals or radionuclides. The classical remediation method is based on the pumping of groundwater. In some cases the efficiency of metal removal by pumping can be improved by using additives to increase the water solubility of the pollutants. In this way, the number of aquifer volumes that need to be pumped can be reduced. Different methods exist for treatment of pumped-up metal-contaminated water. This paper focused on biological methods, and sulfate reduction followed by metal sulfide precipitation is especially presented as a very efficient technology. Another technology is based on metal adsorption and precipitation properties of several bacteria. This principle was used in a moving bed reactor to allow a process of biomass withdrawal at a rate depending on biomass growth. This active bacterial process allowed a continuous growth of biomass, followed by metal binding, metal precipitation and heavy metal loaded biomass removal. It was also shown that the metal-loaded biomass could be used in a pyrometallurgical treatment plant. Extensive attention was paid to the development of *in situ* methods to immobilize metals in the aquifer by using the sulfate reducing capacity of the aquifer microbial biomass leading to metal sulfide bioprecipitation. We described the different parameters that are important to keep the processes running and sustainable (stable precipitates). Finally some typical combinations of processes were presented, i.e. the removal of radionuclides from AMD by sulfate reduction compared with oxidation and adsorption processes or the removal of heavy metals together with boron and barium by a combination of sulfate reduction and adsorption processes. This paper furthermore focused on the combination of the ISBP-process with other techniques and dedicated special attention to the longevity and sustainability of the process. The parameters pH, redox potential, and sulfate concentration seem to play a very important role in the induction of the sulfate reduction process. The selection of the electron donor is important to maintain the process. Combinations of electron donors may turn out to give best results, as suggested by Agathos (2005) for anaerobic dehalogenation. They suggested induction of the process with lactate to stimulate a large variety of dehalogenating bacteria. Later on, the process could be kept going by adding much cheaper methanol. Some experiments also reveal the importance of certain trace elements that may be necessary in order to render the process sustainable. This was shown by some initial results that indicated that the addition of a nitrogen and phosphorus source to lactate kept the process much more stable. It is possible that other metals or elements can become limiting too. However, the mineral aquifer can probably be a source of slowly released micro-elements. Another point that needs evaluation is the comparison between the regular injection of electron donor and slow-release based processes. In fact, all column experiments in the present study were based on simultaneous addition of the electron donor with the groundwater, so the effects of slow release versus immediate injection could not be seen. The regular injection of an electron donor has some drawbacks as it is laborious and even when fully automatically operated needs maintenance. A regular injection of a carbon source in an injection or monitoring well leads to risks related to bacterial growth on the housing of the injection well which can lead to biofouling and blocking of the filters. On the other hand, slow release compounds are more expensive but need only to be injected once or twice per year by a direct-push system (e.g. a Geoprobe system) without risk of clogging or biofouling.

As many sites are contaminated by several metals (anions and cations or radionuclides) by activities such as non-ferrous metals processing, surface treatment and mining activity (including coal mining in some cases), pump and treat methods cannot always deliver an economically acceptable solution. Therefore the development of *in situ* technology is becoming increasingly important. The results and problems encountered up to now indicate that in many cases the combination of different passive systems (also in combination with wetlands) will be necessary (a detailed account of wetlands has not been included in this review). The use of *in situ* bioremediation processes combined or not with other passive systems such as wetlands, PRB, etc. will allow an economically acceptable management of risks related to metals and acid spreading into the environment from large contaminated sites. Pump and treat technology can be useful in source-removal whereas the passive treatment systems will be more applicable for the treatment of diffuse pollution of large affected sites.

Acknowledgements

The research described was made possible by the EU-fifth framework project METALBIOREDUCTION (EVK1-CT-1999-00033) and WELCOME (EVK1-2001-00132) and based on several feasibility studies carried out for several industrial partners.

References

Agathos, S. (2005). Reible, D. D. and Lanczos, T., Ed. NATO Science Series IV, Earth and *Environmental Sciences*, Vol. 73. Springer-Verlag, Dordrecht, Netherlands. Assessment and Remediation of Contaminated Sediments, NATO-ARW Course on Sediment risks and treatment, Bratislava, 16–21 May 2005.

Annachhatre, A. P. and Suktrakoolvait, S. (2001). Biological sulfate reduction using molasses as a carbon source, *Water Environ. Res.*, 73, issue 1, 118-126. ISSN: 1061-4303, WEF (= Water Environment Federation), Alexandria, VA , USA.

Bastiaens, L., Gemoets, J., Van Linden, J., Vermeiren, G., Peeters, P., Boënne, W., Luyckx, J. P., and Diels, L. (2005). Alkalinity generating PRB for in situ treatment of a low pH groundwater. In Permeable Reactive Barriers & Reactive Zones, Proceedings of the 2nd International PRB/RZ symposium, Antwerp, November 14–16, edited by Bastiaens, L., ISBN905857007x, 78.

Benner, S. G., Blowes, D. W. and Ptacek, C. J. (1997). A full-scale porous reactive wall for prevention of acid mine drainage. *Ground water monitoring & remediation*, 17, pp 99-107.

Cameron, R. E. (1992). Guide to site and soil description of hazardous waste site characterization. Volume 1: Metals. Environmental Protection Agency EPA/600/4-91/029.

Diels, L., Spaans, P. H., Van Roy, S., Hooyberghs, L., Wouters, H., Walter, E., Winters, J., Macaskie, L., Pernfuss, B., Woebking, H., and Pümpel, T. (2001). Heavy metals removal by sand filters inoculated with metal sorbing and precipitating bacteria. Biohydrometallurgy - Fundamentals, Technology and Sustainable Development (part B) - pp 317-326.

Diels, L., Geets, J., Dejonghe, W., Van Roy, S. and Vanbroekhoven, K. (2005a). Heavy metal immobilization in groundwater by in situ bioprecipitation: comments and questions about carbon source use, efficiency and sustainability of the process. Consoil 2005.

Diels, L., Geets, J., Van Roy, S., Dejonghe, W., Gemoets, J. and Vanbroekhoven, K. (2005b). Bioremediation of heavy metal contaminated sites. In Soil Remediation series 6 (Proceedings of the European Summer School on Innovative Approaches to the Bioremediation of Contaminated Sites), ed. Faba, F., Canepa, P.

Diels, L., Spaans, P. H., Van Roy, S., Hooyberghs, L., Wouters, H., Walter, E., Winters, J., Macaskie, L., Finlay, J., Pernfuss, B., Woebking, H. and Pümpel, T. (2003). Heavy metals removal by sand filters inoculated with metal sorbing and precipitating bacteria. *Hydrometallurgy*, 71, pp 235-241.

El Fantroussi, S., Naveau, H., and Agathos, S. N. (1998). Anaerobic dechlorinating bacteria. *Biotechnol. Prog.*, 14, pp 167-175.

Francis, A. J. and Dodge, C. J. (1993). Influence of complex structure on the biodegradation of iron-citrate complexes. *Appl. Environ. Microbiol.*, 59, pp 109-113.

Fürst, P. and Burggräf, H. (2000). Das Bio-Substrat Verfahren firmenbericht Dr. Fürst systems.

Geets, J., Borremans, B., Vangronsveld, J., Diels, L. and van der Lelie, D. (2004). Molecular monitoring of SRB community structure and dynamics in batch experiments to examine the applicability of in situ precipitation of heavy metals for groundwater remediation. *J. Soil Science*, p 1-15 (OnlineFirst), Ecomed Publishers (Landsberg, Germany).

Ghyoot, W., Feyaerts, K., Diels, L., Vanbroekhoven, K., de Clerck, X., Gevaerts, G., Ten Brummeler, E. and van den Broek, P. (2004). In situ bioprecipitation for remediation of metal-contaminated groundwater. Edited by W. Verstraete, Published by Elsevier. In: European Symposium on Environmental Biotechnology, ESEB 2004. p 241-244. ISBN 90 5809 653 X.

Gilbert, O., de Pablo, J., Cortina, J. L. and Ayora, C. (2002). Treatment of acid mine drainage by sulfate-reducing bacteria using permeable reactive barriers: a review from laboratory to full-scale experiments. Reviews in *Environmental Science and Biotechnology*, 1, pp 327-333.

Greben, H. A., Maree, J. P., Singmin, I. and Mnqanqeni, S. (2000). Biological sulphate removal from acid mine effluent using ethanol as carbon and energy source. *Water Sci. Technol.*, 42, pp 339-344.

Grimault, J. O., Ferer, M. and Macpherson, E. (1999). The mine tailing accident in Aznacollar. *The Science of the Total Environment*, 242, pp 3-11.

Groudev, S., Georgiev, P. S., Spasova, I. I., Nicolova, M. V. and Diels, L. (2004). Bioremediation of acid drainage by means of a passive treatment system. The 20th Annual International Conference on Soils, Sediments and Water, Amherst (USA), October 18-21.

Groudev, S. N., Nicolova, M. V., Spasova, I. I., Groudeva, V. I., Georgiev, P. S. and Diels, L. (2005a). Bioremediation of acid drainage by means of a passive treatment system. Proceedings of 16th International Biohydrometallurgy Symposium, Cape Town, 25-29 September, pp 473-478 (CD-rom version), ISBN 1-920051-17-1, Comgress (Cape Town, South Africa).

Groudev, S. N., Spasova, I. I., Georgiev, P. S. and Nicolova, M. V. (2005b). Bioremediation of acid drainage in a uranium deposit. The fourteenth Annual West Coast Conference on Soils, Sediments and Water, San Diego (USA), March 15-18, p 66, AEHS (USA).

Hammack, R. W. and Edenborn, H. M. (1992). The removal of nickel from mine waters using bacterial sulfate reduction. *Appl. Microbiol. Biotechnol.*, 37, pp 674-678.

Hao, O. L. (2000). In Environmental technologies to treat sulfur pollution: principles and engineering. Ed. Lens, P. and Hulshoff-Pol, L., IWA Publishing, London, pp 393-414.

Janssen, G. M. C. M. and Temminghoff, E. J. M. (2004). In situ metal precipitation in a zinc-contaminated, aerobic sandy aquifer by means of biological sulfate reduction. *Environ. Sci. Technol.*, 38, pp 4002-4011.

Johnson, B. D. and Hallberg, K. N. (2005). Acid mine drainage remediation options: a review. *Science of the Total Environment*, 338, pp 3-14.

Koenigsberg, S. S. (2002). Metals remediation compound, www.regenesis.com.

Koenigsberg, S. S., Sandefur, C. A., Lapus, K. A. and Pasrich, G. P. (2002). Facilitated desorption and incomplete dechlorination observations from 350 applications of HRC. www.regenesis.com.

Kontopoulos, A. (1988). Acid mine drainage control. In Effluent Treatment in the Mining Industry. Ed. Castro, S. H., Vergara, F. and Sánches, M. A., Concepción, pp 57-118.

Mallants, D., Diels, L., Bastiaens, L., Vos, J., Moors, H., Wang, L., Maes, N. and Vandenhove, H. (2002). Removal of uranium and arsenic from groundwater using six different reactive materials: assessment of removal efficiency. Int. Conf. Uranium Mining and Hydrogeology UMH III, Freiburg, (Germany), 15-21/09/2002. Ed. Merkel, B. J., Planer-Friedrich, P., Wolkersdorfer, C. pp 565-571.

Maree, J. P. and Strydom, W. W. (1987). Biological sulphate removal from industrial effluent in an upflow packed bed reactor. *Water Res.*, 21, pp 141-146.

Merten, D., Kothe, E. and Büchel, G. (2004). Studies on microbial heavy metal retention from uranium mine drainage water with special emphasis on rare earth elements. *Mine Water and the Environment*, 23, pp 34-43.

Mulligan, C. N., Yong, R. N. and Gibbs, B. F. (2001). Remediation technologies for metal-contaminated soils and groundwater: an evaluation. *Engineering Geology*, 60, pp 193-207.

Munro, L. D., Clark, M. W. and McConchie, D. (2004). A Bauxsol™-based permeable reactive barrier for the treatment of acid rock drainage. *Mine Water and the Environment*, 23, pp 183-194.

Patterson, J. W. (1985). Industrial Wastewater Treatment Technology, 2nd ed. Butterworth, Boston.

Powell, R. M., Puls, R. W., Hightower, S. K. and Sabatini, D. A. (1995). Coupled iron corrosion and chromate reduction: mechanisms for subsurface remediation. *Environ. Sci. Technol.*, 29, pp 1913-1922.

Prasad, D., Wai, M., Berube, P. and Henry, J. G. (1999). Evaluating substrates in the biological treatment of acid mine drainage. *Environ. Technol.*, 20, pp 449-458.

Pümpel, T., Ebner, C., Pernfuss, B., Schinner, F., Diels, L., Keszthelyi, Z., Macaskie, L., Tsezos, M. and Wouters, H. (2001a). Removal of nickel from plating rinsing water by a moving-bed sandfilter inoculated with metal sorbing and precipitating bacteria. *Hydrometallurgy*, 59, pp 383-393.

Pümpel, T. and Paknikar, K. (2001b). Bioremediation technologies for metal-containing wastewaters using metabolically active microorganisms. *Advances in Applied Microbiology*, 48, pp 135-169.

Saha, G. C. and Ali, M. A. (2001). Groundwater contamination in the Dhaka city from tannery industry. *Journal of Civil Engineering*, CE 29. No. 2. The Institution of Engineers, Bangladesh.

Schmidt, E. (2002). Actief Bodembeheer de Kempen (ABdK). Raamplan Actief Bodembeheer de Kempen 2002–2004.

Steketee, J. (2004). Beslissingsondersteunend systeem vastlegging van zware metalen in de verzadigde zone van de bodem. SKB-report SV-615.

Temminghoff, E. and Janssen, G. (2005). PAO-cursus Natuurlijke en gestimuleerde vastlegging van zware metalen in de bodem, Delft, 8-9 March 2005.

Tessier, A., Campbell, P. G. C. and Bisson, M. (1979). Sequential extraction procedure for the speciation of particulate trace metals. *Anal. Chem.* 51, pp 844-851.

Van Houten, R. T., Hulshoff-Pol, L. W. and Lettinga, G. (1994). Biological sulfate reduction using gas-lift reactors fed with hydrogen and carbon-dioxide as energy and carbon source. *Biotechnol. Bioeng.*, 44, pp 586-594.

Van Roy, S., Bastiaens, L., Vanbroekhoven, K., Dejonghe, W. and Diels, L. (2005a). Sorbent screening for in situ treatment of groundwater with heavy metals. The eighth In situ and On-site Bioremediation Symposium, June 6-9, 2005, Baltimore, E-21 (CD-rom version), ISNB 1-57477-152-3, Batelle Press (Columbus, USA).

Van Roy, S., Vanbroekhoven, K. and Diels, L. (2005b). Immobilisation of heavy metals in the saturated zone by sorption and in situ bioprecipitation processes. Proceedings of the IMWA meeting in Oviedo (5-7 September). Editors Loredo, J., Pendas, F., ISBN 84-689-3415-1, pp 355-360, University of Oveido (Oveido, Spain).

Vanbroekhoven, K., Geets, J., Van Roy, S. and Diels, L. (2005a). Impact of DOC on precipitation and stability of metal sulfides during evaluation of ISBP in column experiments, Proceedings of Consoil 2005, October 3-7 2005, Bordeaux (France), pp 1875-1879 (CD-rom version), ISBN 3-923704-50-x, F&U Confirm (Germany).

Vanbroekhoven, K., Dejonghe, W., Nuyts, G. and Diels, L. (2005b). Feasibility study of the NA potential for mixed plumes: impact of heavy metals on halorespiration. The eighth In situ and On-site Bioremediation Symposium, June 6-9, 2005, Baltimore. Submitted.

Vanbroekhoven, K., Ryngaert, A., Van Roy, S., Diels, L. and Dejonghe, W. (2006). Competitive dissimilatory iron reduction during in situ bioprecipitation of metals: Evidence by quantitative PCR using SRB and Geobacter specific primers. ESEB 2006, Leipzig, in preparation.

Wagner-Döbler, I., von Canstein, H., Li, Y., Timmis, K. N. and Deckwer, W. D. (2000). Removal of Mercury from chemical wastewater by microoganisms in technical scale, *Environ. Sci. Technol.*, Vol. 43, Issue 21, pp 4628-4634. DOI: 10.1021/es0000652.

Waybrant, K. R., Ptacek, C. J. and Blowes, D. W. (2002). Treatment of mine drainage using permeable reactive barriers: column experiments. *Environ. Sci. Technol.*, 36, pp 1349-1356.

Webb, J. S., McGinness, S. and Lappin-Scott, H. M. (1998). Metal removal by sulfate-reducing bacteria from natural and constructed wetlands. *J. Appl. Microbiol.*, 84, pp 240-248.

Weijma, J., Copini, C. F. M., Buisman, C. J. N., and Schultz, C. E. (2002). Biological recovery of metals, sulfur and water in the mining and metallurgical industry. In *Water Recycling and Recovery in Industry*/Lens, P. N. L., Hulshoff-Pol, L. W., Wilderer, P., Asano, T. pp 605-622. London, UK: IWA Publishing.

Woebking, H. and Diels, L. (2000). Abreicherung und Rückgewinnung von Eisen und Nichteisenmetallen aus industriellen Abwässeren unter Verwendung eines Bakterien geimpften Sandfilters. *Berg- und Hüttenmännische Monatshefte* 7, pp 265-270.

BIOREMEDIATION IN SITU OF POLLUTED SOIL IN A URANIUM DEPOSIT

STOYAN GROUDEV, IRENA SPASOVA, MARINA NICOLOVA AND PLAMEN GEORGIEV
University of Mining and Geology "Saint Ivan Rilski", Studentski grad, Sofia 1700, Bulgaria

Abstract. An experimental plot containing acidic soil heavily contaminated with radionuclides (mainly uranium) and heavy metals (mainly copper, zinc and cadmium) was treated by means of an in situ biotechnological method based on the activity of the indigenous microflora, mainly of some acidophilic chemolithotrophic bacteria. The pollutants were located mainly in the upper soil layers (horizon A) and considerable portions of them were present in forms susceptible to bacterial leaching. The treatment was connected with the dissolution of pollutants and their removal from the soil profile by means of water acidified with sulphuric acid to pH of about 3.5. The bacterial activity was enhanced by suitable changes of some essential environmental factors such as pH and water, oxygen and nutrient contents in the soil. The removal of pollutants was very efficient and within 24 months their residual concentrations in the soil were decreased below the relevant permissible levels. The pregnant soil effluents containing the dissolved pollutants were efficiently cleaned up by wetlands located near the experimental plot.

Keywords: soil bioremediation, radionuclides, heavy metals, chemolithotrophic bacteria, wetlands

1. Introduction

The uranium deposit Curilo, located in Western Bulgaria, for a long period of time was a site of intensive mining activities including both the open-pit and underground techniques as well as *in situ* leaching of uranium. The mining operations were ended in 1990 but until now both the surface and ground waters and soils within and near the deposit are heavily polluted with radionuclides (mainly uranium) and heavy metals (mainly copper, zinc and cadmium).

M.D. Annable et al. (eds.), Methods and Techniques for Cleaning-up Contaminated Sites, 25–34.

Laboratory experiments carried out with soil samples from the deposit revealed that an efficient remediation of the soils was achieved by solubilizing the pollutants and washing the soil profile by means of acidified water solutions. The solubilization was connected with the activity of the indigenous soil microflora, mainly with the activity of some acidophilic chemolithotrophic bacteria. It was possible to enhance considerably this activity by suitable changes in the levels of some essential environmental factors such as pH and waters, oxygen and nutrient contents of the soil.

The promising data from the laboratory experiments as well as from the geologic and hydrogeologic investigations of the polluted lands were the main reason for the application of the above-mentioned soil treatment *in situ* under real field conditions in an experimental plot located in the deposit. Some data from these field experiments are shown in this paper.

2. Materials and Methods

A detailed sampling procedure was carried out to characterize the soil and the subsurface geologic and hydrogeologic conditions. Surface and bulk soil samples up to a depth of 2 m were collected by an excavator. Drill hole samples were collected up to a depth of 8 m. Elemental analysis in the samples was performed by digestion and measurement of the ion concentration in solution by atomic absorption spectrophotometry and induced coupled plasma spectrophotometry. Mineralogical analysis was carried out by X-ray diffraction techniques.

The main geotechnical characteristics of the site, such as permeability and wet bulk density, were measured *in situ* using the sand-core method (U.S. Environmental Protection Agency, 1991). True density measurements were carried out in the laboratory using undisturbed core samples. Such samples were also used for determination of their acid generation and net neutralization potentials using static acid-base accounting tests. The bioavailable fractions of the pollutants were determined by leaching the samples with DTPA and EDTA (Sobek et al., 1978). The mobility of the pollutants was determined by the sequential extraction procedure (Tessier et al., 1979). The toxicity of soil samples was determined by the EPA Toxicity Characteristics Leaching Procedure (U.S. Environmental Protection Agency, 1990).

The experimental plot had a rectangular shape and was 80 m^2 in size (10 m × 8 m). Water acidified with sulphuric acid to pH of about 3.5 was used as leach solution. The upper soil layers were ploughed up periodically to enhance the natural aeration.

The flow sheet of the operation included also a system to collect the soil drainage solutions and to avoid their seepage and the distribution of contaminants into the environment. The system consisted of several ditches and wells located in suitable sites in the experimental plot. The soil effluents collected by this system were then treated initially by a natural and since the beginning of the second year by a constructed wetland located near the experimental plot to remove the dissolved contaminants. These wetlands were characterized by an abundant water and emergent vegetation and a diverse microflora. *Typha latifolia, Typha angustifolia, Phragmites australis* and different algae were the prevalent plant species in the wetland.

A plot with the same size and shape was used as a control. This plot also consisted of polluted soil. This soil, however, was not treated during the whole experimental period.

The isolation, identification and enumeration of microorganisms were carried out by methods described elsewhere (Karavaiko et al., 1988; Groudeva et al., 1993).

The bacterial activity *in situ* in the soil was determined by following the rates of ferrous iron oxidation in samples of drainage waters collected from different sections of the experimental and control plots as well as in 9K nutrient medium (Silverman and Lundgren, 1999) inoculated with freshly collected soil samples. These experiments were carried out in 300 ml Erlenmeyer flasks containing 100 ml liquid phase. The flasks were incubated *in situ*, at different depths in the plots, at the relevant natural temperatures, for 5 days. The technique described by Karavaiko and Moshniakova (1971) was used with some modifications (Groudev and Groudeva, 1993) to determine the $^{14}CO_2$ fixation *in situ*.

3. Results and Discussion

The soil profile was approximately 90 cm deep (horizon A, 30 cm; horizon B, 40 cm; horizon C, 20 cm). The soil profile was underlined by intrusive rocks with a very low filtration coefficient (approximately 7×10^{-8} m/s).

The groundwater level was located at about 10 m below the surface. The surface and groundwaters in the site were separated by the impermeable rocks.

Data about the chemical composition and some essential geotechnical parameters of the soil are shown in Tables 1 and 2. The concentrations of contaminants were higher in the upper soil layers (in the horizon A). Considerable portions of the contaminants were present in fractions susceptible to biological leaching and within the experimental period of 24 months their residual concentrations in the horizon A were decreased below the relevant permissible levels.

At the same time, the decrease in the concentrations of contaminants in the control plot as a result of the processes of natural attenuation was negligible.

The analysis of the microflora in the experimental plot showed that it contained a rich variety of microorganisms (Table 3). The mesophilic acidophilic chemolithotrophic bacteria related to the species *Acidithiobacillus ferrooxidans, Acidithiobacillus thiooxidans* and *Leptospirillum ferrooxidans* were the prevalent microorganisms in the top soil layers. These bacteria were able to oxidize the sulphide minerals present in the soil and to solubilize their metal components. The non-ferrous metals, i.e. copper, zinc, cadmium, were removed mainly in this way. Uranium was also solubilized as a result of its prior bacterial oxidation from the tetravalent to the hexavalent state. The activity of these acidophilic chemolithotrophic bacteria *in situ* was quite high (Table 4). Thermophilic chemolithotrophic bacteria able to oxidize sulphide minerals and uranium were not found in the soil. However, some basophilic chemolithotrophic species (mainly *Thiobacillus thioparus, T. denitrificans* and *Halothiobacillus neapolitanus*) as well as some acidophilic heterotrophs (mainly such related to the genus *Acidiphilium*) were also present but in lower numbers. In the deeply located soil layers (in the subhorizon B_2) both the total number and diversity of the microorganisms were much lower.

TABLE 1. Characteristics of the soil treated in this study

Parameters	Horizon A (0 – 30 cm)	Horizon B (31 – 70 cm)
Chemical composition (in %):		
- SiO_2	64.4	67.5
- Al_2O_3	12.2	14.1
- Fe_2O_3	10.4	9.1
- CaO	0.41	0.37
- MgO	0.44	0.35
- K_2O	1.45	1.27
- S total	1.04	0.80
- S sulphidic	0.86	0.75
- Humus	2.1	0.8
Bulk density, g/cm^3	1.22	1.40
Specific density, g/cm^3	2.80	3.02
Porosity, %	51	46
Moisture capacity, %	48	44
Permeability, cm/s	8×10^{-2}	6×10^{-2}
pH (H_2O)	4.6	4.8
Net neutralization potential, kg $CaCO_3$/t	- 170	- 134

TABLE 2. Content of contaminants in the horizon A of the soil before and after the treatment

Parameters	Cu	Zn	Cd	U
Content of contaminants, ppm				
- before treatment	170	194	4.1	59
- after treatment	32	44	0.9	8.2
Permissible levels for soils with pH 4.1 – 5.0	40	60	1.5	10
Permissible levels for soils with pH <4.0	20	30	0.5	10
Bioavailable fraction, ppm:				
a. by DTPA leaching				
- before treatment	28	25	0.7	9.1
- after treatment	3.7	3.2	0.05	2.8
b. by EDTA leaching				
- before treatment	23	14	0.4	4.6
- after treatment	1.5	1.4	0.01	0.8
Easily leachable fractions – exchangeable + carbonate, ppm				
- before treatment	64	71	1.0	17
- after treatment	3.5	4.4	0.03	2.1
Inert fraction, ppm				
- before treatment	71	82	2.1	14
- after treatment	27	37	0.8	6.4
Pollutants solubilized during the toxicity test, ppm				
- before treatment	5.94	7.52	0.14	1.35
- after treatment	0.35	0.41	0.01	0.17

TABLE 3. Microorganisms in the horizon A of the experimental plot before and after the treatment

Microorganisms	Before treatment	During treatment
	Cells/g dry soil	
Aerobic heterotrophic bacteria	$10^3 - 10^7$	$10^3 - 10^6$
Fe^{2+} - oxidizing chemolithotrophs (at pH 2.5)	$10^1 - 10^3$	$10^4 - 10^7$
S^0 - oxidizing chemolithotrophs (at pH 2.5)	$10^2 - 10^4$	$10^4 - 10^7$
$S_2O_3^{2-}$ - oxidizing chemolithotrophs (at pH 7)	$10^2 - 10^5$	$10^1 - 10^4$
Nitrifying bacteria	$10^1 - 10^4$	$< 10^2$
Nitrogen – fixing bacteria	$10^2 - 10^4$	$< 10^2$
Streptomycetes	$10^2 - 10^5$	$10^1 - 10^4$
Fungi	$10^2 - 10^6$	$10^1 - 10^4$
Anaerobic heterotrophic bacteria	$10^3 - 10^5$	$10^2 - 10^4$

TABLE 4. Microbial activity in situ at different environmental conditions

Sample tested	Fe^{2+} oxidized for 5 days, g/l	$^{14}CO_2$ fixed for 5 days, counts/min.ml (g)
Soil effluents with a pH of 4.6 + Fe^{2+} (9 g/l) at 8 – 10°C	0.46 – 1.32	1400 – 3500
Soil effluents with a pH of 3.7 + Fe^{2+} (9 g/l) at 8 – 10°C	0.86 – 2.75	2100 – 6400
Soil effluents with a pH of 3.7 + Fe^{2+} (9 g/l) at 17 – 20°C	1.45 – 4.10	3700 – 10400
Ore suspensions in 9K nutrient medium (with 9 g/l Fe^{2+} and pH 3.7) at 8 – 10°C	0.55 – 1.59	1500 – 4100
Ore suspensions in 9K nutrient medium (with 9 g/l Fe^{2+} and pH 3.7) at 17 – 20°C	0.95 – 2.99	2300 – 7100
Ore suspensions in 9K nutrient medium (with 9 g/l Fe^{2+} and pH 4.6) at 17 – 20°C	0.68 – 2.08	1700 – 5100

The number and activity of the acidophilic chemolithotrophs in the soil were limited by some essential environmental factors such as the relatively high soil pH, shortage of oxygen and of some important nutrients such as nitrogen and phosphorus sources in the soil, insufficient soil moisture during relatively long periods of time. The treatment of the contaminated soil was connected with the increasing number and activity of the indigenous acidophilic chemolithotrophic microorganisms by suitable changes in the levels of the above-mentioned environmental factors. This was achieved by regular ploughing up and irrigation of the soil and by addition of some essential nutrients. The optimum soil humidity was about 50% of the moisture capacity of the soil, but periodic flushing with slightly acidified water (with pH of about 3.5) was needed to remove the soil contaminants. Zeolite saturated with ammonium phosphate was added to the soil (in amounts of 5 kg/t dry soil) to provide the microorganisms with ammonium and phosphate ions to improve the physico-mechanical properties of the soil.

The rate of soil clean-up markedly depended on the temperature. The highest rates were achieved during the warm summer months at soil temperatures, which exceeded 25°C. It must be noted, however, that the clean-up was efficient even at temperatures as low as 5 – 10°C, and practically stopped only during the cold winter months (December – February) when the temperatures were often close or even below 0°C.

The soil in the control plot was characterized by much lower content and activity of microorganisms.

Some amounts of contaminants solubilized in the upper soil horizon A were retained in the deeper horizons, mainly due to processes such as the microbial dissimilatory sulphate reduction and sorption on the clay minerals.

The population of sulphate-reducing bacteria in these deeply located soil layers was quite diverse but not numerous during the treatment (Table 5). This was due to the acidic pH and relatively low concentration of dissolved organic compounds, which were used as sources of energy and carbon by these bacteria. At the same time, the contents of pollutants in the deeply located soil layers were also decreased as a result of bioleaching processes carried out by different microorganisms, mainly by the chemolithotrophic sulphide-oxidizing acidophilic bacteria and the products from their oxidative activity (ferric ions and sulphuric acid). Some role was played also by the basophilic chemolithotrophs *H. neapolitanus* and *T. thioparus*, which were found in some microzrones with pH level of about 3 – 3.5, i.e. at the limit allowing the growth and activity of these bacteria. They oxidized the elemental sulphur formed as a result of different chemical, electrochemical and biological processes and deposited on the surface of the sulphides. It must be noted that both the acidophilic and basophilic chemolithotrophs in deeply located soil layers were active at microaerophilic conditions created by the oxygen transported to these layers by means of the drainage waters.

Some anaerobic heterotrophic bacteria, mainly such possessing iron and manganese respiration, were able to solubilize iron and manganese hydroxide and oxide minerals and in this way to liberate the non-ferrous metals and uranium encapsulated in these minerals. Portions of the non-ferrous metals were solubilized as complexes with some organic compounds. Most of the lead was solubilized in this way.

TABLE 5. Sulphate-reducing bacteria in the soil subhorizon B_2 during the treatment

Sulphate-reducing bacteria	Cells/ml pore solution
Desulfovibrio (mainly *D. desulfuricans*)	$10^1 - 10^3$
Desulfobulbus (mainly *D. elongatus*)	$10^1 - 10^3$
Desulfococcus (mainly *D. postgatei*)	$0 - 10^2$
Desulfobacter (D. multivorans)	$0 - 10^2$
Desulfotomaculum (mainly *D. nigrificans*)	$0 - 10^1$
Desulfosarcina (D. variabilis)	$10^1 - 10^2$
Desulfomonas (non-identified species)	$0 - 10^1$

As a result of the microbial activity not only the horizon A but, to some extent, the whole soil profile was efficiently cleaned at the end of the experimental period from the easily leachable forms of the pollutants. Regardless of the precipitation of portions of the pollutants in the subhorizon B_2, the soil effluents usually still contained most pollutants in concentrations higher than the relevant permissible levels for water intended for use in the agriculture and/or industry. These effluents were efficiently treated by means of the wetlands located near the experimental plot (Table 6). The removal of contaminants in these wetlands was connected with different processes but the microbial sulphate reduction and the sorption of contaminants on the organic matter (mainly living and dead plant biomass) and clay minerals present in the wetlands played the main role. Portions of iron and manganese were removed as a result of the prior oxidation of Fe^{2+} and Mn^{2+} to Fe^{3+} and Mn^{4+}, respectively, followed by precipitation of these higher valency forms as $Fe(OH)_3$ and MnO_2. Data about the microflora of the soil effluents and wetlands are shown in Table 7.

TABLE 6. Composition of the soil effluents before and after treatment in the wetlands

Parameters	Before treatment	After treatment	Permissible levels for waters used in agriculture or industry
pH	3.0 – 3.7	7.1 – 7.7	6 – 9
Eh, mV	(-95) – (-203)	(+235) – (+307)	–
Dissolved oxygen, mg/l	0.3 – 0.9	2.1 – 5.0	2
Total dissolved solids, mg/l	482 – 1294	275 – 655	1500
Solids, mg/l	38 – 82	25 – 73	100
Dissolved organic carbon, mg/l	14 – 41	17 – 46	20
Sulphates, mg/l	230 – 648	140 – 320	400
Uranium, mg/l	0.05 – 0.88	< 0.05	0.6
Copper, mg/l	0.35 – 2.42	< 0.2	0.5
Zinc, mg/l	0.41 – 9.50	0.05 – 0.91	10
Cadmium, mg/l	< 0.01 – 0.07	< 0.01	0.02
Lead, mg/l	0.15 – 1.22	< 0.1	0.2
Iron, mg/l	15 – 38	0.7 – 2.3	5
Manganese, mg/l	4.1 – 18.1	< 0.5	0.8

The chemical composition, structure and main physical and water properties of the soil after the treatment were altered to a small extent. However, the pH of the soil was decreased from the initial 4.6 to 3.5. After the two-year treatment period the experimental plot was subjected to some conventional remediation procedures such as grassing of the treated soil, addition of suitable fertilizers and animal manure as well as with periodical ploughing up, liming (to increase and maintain the pH to about 5) and irrigation. As a result of this, the quality of the soil was completely restored. No soluble forms of the above-mentioned contaminants in concentrations higher than the relevant permissible levels were detected so far (approximately five years after the end of the treatment) in the soil pore and drainage waters after rainfall.

TABLE 7. Microflora of the fresh soil effluents and the constructed wetland

Microorganisms	Samples		
	Soil effluents before treatment	Waters from the wetland	Sediments from the wetland
		Cells/ml (g)	
Aerobic heterotrophic bacteria	$10^1 - 10^3$	$10^2 - 10^6$	$10^1 - 10^4$
Fe^{2+} - oxidizing chemolithotrophs (at pH 2.5)	$10^3 - 10^5$	$0 - 10^1$	0
$S_2O_3^{2-}$ - oxidizing chemolithotrophs (at pH 7)	$10^1 - 10^3$	$0 - 10^2$	$0 - 10^1$
Fe^{2+} - oxidizing heterotrophs (at pH 7)	$0 - 10^2$	$10^2 - 10^4$	$0 - 10^2$
Cellulose-degrading aerobes	$0 - 10^1$	$0 - 10^4$	$0 - 10^2$
Anaerobic heterotrophic bacteria	$10^1 - 10^4$	$10^2 - 10^4$	$10^2 - 10^5$
Fe^{3+} - reducing bacteria	$10^1 - 10^3$	$0 - 10^3$	$10^2 - 10^5$
Sulphate-reducing bacteria	$10^1 - 10^3$	$0 - 10^2$	$10^2 - 10^4$
Cellulose-degrading anaerobes	$0 - 10^1$	$0 - 10^2$	$10^1 - 10^4$

Acknowledgement

A part of this work was financially supported by the National Fund "Scientific researches" under the project CENBIOHEALTH.

References

Groudev, S.N. and V.I. Groudeva, 1993. "Microbial Communities in Four Industrial Copper Dump Leaching Operations in Bulgaria", FEMS Microbiological Reviews, 11:261–268.

Groudeva, V.I., I.A. Ivanova, S.N. Groudev and G.C. Uzunov. 1993. "Enhanced Oil Recovery by Stimulating the Activity of the Indigenous Microflora of Soil Reservoirs". In A.E. Torma, M.L. Appel and C.L. Brierley (Eds.), Biohydrometallurgical Technologies, vol. II, pp. 349-356. TMS Minerals, Metals & Materials Society, Warrendale, PA.

Karavaiko, G.I. and S.A. Moshniakova, 1971. "A Study on Chemosynthesis and Rate of Bacterial and Chemical Oxidative Processes Under Conditions of Copper-nickel Ore Deposits of Kolsky Peninsula", Mikrobiologiya, 40:551–557 (in Russian).

Karavaiko, G.I., G. Rossi, A.D. Agate, S.N. Groudev and Z.A. Avakyan, (Eds.) 1988. Biotechnology of Metals. Manual, Center for International Projects GKNT, Moscow.

Silverman, M.P. and D.G. Lundgren, 1959. "Studies on the Chemoauthotrophic Iron Bacterium Ferrobacillus ferrooxidans. I. An improved medium and a harvesting procedure for securing high cell yields", Journal of Bacteriology, 77:642–647.

Sobek, A.A., W.A. Schuller, J.R. and R.M. Smith, 1978. Field and Laboratory Methods Applicable to Overburden and Mine Soils, US Environmental Protection Agency, Report 600/2-78-054.

Tessier, A., P.G.C. Campbell and M. Bisson, 1979. "Sequential Extraction Procedure for Speciation of Particulate Trace Metals", Analytical Chemistry, 51(7):844–851.

U.S. Environmental Agency, 1990, Characteristics of EP Toxicity, Paragraph 261.24, Federal Register 45 (98).

U.S. Environmental Protection Agency, 1991. Description and Sampling of Contaminated Soils – A Field Pocket Guide. EPA/625/12-91/002 Technology Transfer, Centre for Environmental Research Information, U.S. Environmental Protection Agency, Cincinnati, OH.

VITAL SOIL VERSUS CONTAMINANTS

P. DOELMAN
Doelman Advice, August Faliseweg 10, 6703 AS Wageningen, the Netherlands p.doelman@chello.nl

1. Introduction

The vital soil gradually evolved after the formation of earth. Beyond the illusion of the day there should be awareness that recycling of elements is a phenomenon existing over billions of years. Whereas the anoxic processes are far older than the oxic ones and consequently have a longer history of evolution. The principle of the infallibility of nature in recycling of elements is also applicable for the biodegradation of organic contaminants.

The history of environmental pollution, resulting in soil/sediment contamination, is relatively short. One of the first pollution affairs dates from the Roman period (Hong et al., 1996) and refers to heavy metals, the open smelting of ores. Since than non-optimal metal-processing industries resulted in heavy metal fallout, causing elimination of sensitive species in soils but also adaptation. Metal resistant grasses are Agrostis and Festuca (Ernst 1989), whereas *Arabidopsis hulleri* became a hyper accumulator of Zn (Ernst, 2004). The soil microflora (Doelman et al., 1994) and the soil fauna (Hopkins, 1994) became affected in many ways. The soil contamination by organics is more recent and is mostly the result of mismanagement, such as overdoses of plant-protection chemicals as DDT. Since the early 1950s there is concern on those issues. The book Silent Spring (Carson, 1962) questioned the accumulation of DDT in food chains: birds of prey became well known victims. The relation between soil contamination and higher animals is qualitatively and quantitatively shown by Van den Brink (2004). Simultaneously those spillages became a source of inspiration to study their fate as microbial degradation. The anaerobic degradation pathways of oil compounds such as BTEX (Wilson et al., 1986) and of degreasing compounds, applied in the dry cleaning industry, were discovered and could be applied in in-situ remediation. Degradation rates and preferable degradation conditions became known. In table 1 some relevant historical events are mentioned.

M.D. Annable et al. (eds.), *Methods and Techniques for Cleaning-up Contaminated Sites*, 35–44.

TABLE 1. Some relevant landmarks in the history of bioremediation

Event	Years ago	Reference
Origin of the universe	13.500.000.000	Bryson, 2003
Formation of earth	4.500.000.000	De Duve, 1995
Anaerobic bacteria	3.500.000.000	”
Aerobic bacteria	2.000.000.000	”
Origin eukaryotes	1.500.000.000	”
Cambrian explosion	570.000.000	”
Cretaceous mass extinction	65.000.000	”
Homo sapiens	?: 2.000.000	
Open ore smelters	> 2.000	Hong et al., 1996
Pollution by organics	> 200	
Anaerobic degradation of BTEX	20	Wilson et al., 1986
Halorespiration of VOCl's	15	Holliger, 1992
Dioxine mineralization	2	Bunge et al., 2003
Dehalococcoides, as proven technoloy	0	Bemmel & Klijn, 2006

There is sufficient knowledge available on the recycling of elements, on the biodegradation of organic contaminants and on the ecological recovery of mal-treated soils to come to practical application. There are many lessons to be learned without re-inventing wheels. Whether contamination can be fatal to soil depends on he time period and the degree of contamination, or combinations of contamination, or combined contamination with natural stress. Mostly with a bird's eye view and sometimes in detail, information will be provided on the interaction between contamination and soil with an attempt to reduce the complexity of both to rather simple applicable handles.

2. Contaminants and Biodegradation Principles, Pathways and Rates in Soil

In principle the number of contaminants that has reached soil systems all over the world the last decades is innumerably large. For reasons of simplicity and predicting their behaviour, those "contaminants" can be classified in seven structural groups: heavy metals (H.M), persistent organic pollutants as HCH, HCB, DDT, Dioxine, PCB's, Drins (POP's), crude hydrocarbons and poly aromatic hydrocarbons (PAH's), mono aromatic compounds as Benzene, Toluene, Ethylbenzene and Xylenes (BTEX), chlorinated aliphatics as PER, TRI and chlorinated benzenes, -phenols, etc (VOCl's), Cyanides (CN) and eutrophication elements as nitrogen and phosphorous (N,P). Their behaviour in the process of natural attenuation and the uptake by plants and the soil fauna is qualitatively given in table 2. Natural attenuation is, according to the United States Environmental Protection Agency (USEPA), the process of biodegradation, diffusion, dilution, sorption, volatilisation and chemical stabilization. The scientific literature is loaded with detail information.

TABLE 2. Overview of attenuation processes that affect the persistence of contaminants in soil

	H.M	POP's	PAH's	BTEX	VOCl's	CN
Sorption	++	+++	+++	+	+	+<>+++
Diffusion	+	+	+	++++	+++	+<>+++
Dilution	+	+	+	++++	+++	o<>+++
Evaporation	O	O	O	++++	++	O<>+++
Microbial transformation	+	+ <>++++	o<>++++	+++++	+++	O<>+++++
Chemical stabilization	+	+++	+++	+	+	+++
Uptake plants	++	O	O	O	O	O
Uptake soil fauna	++	++	++	O	O	+

++++: very general and strong
+++: general and strong
++: regularly
+: seldom/hardly
o: not

A Selection of the History of Biodegradation Research

Knowledge on the fate of contaminants in various soil types has also been obtained during decades. Here the attention will mainly be paid to biodegradation. General principles of biodegradation have been published by Alexander (1985) and many others. Atlas (1981) published the general aspects of biodegradation of hydrocarbons (Atlas, 1981). Wilson (1986) and co-workers were one of the first to discover the anaerobic degradation of BTEX and to implement it into in-situ bioremediation of groundwater. The anaerobic degradation of HCH was proven in the laboratory by Bachmann et al. (1988) and in the field by Doelman et al. (1990). For hexachlorocyclohexane (HCH) it took 12 years before the principle of biodegradation was partly unraveled. It should be emphasized that the limitation of biodegradation of organics is often due to non-optimal environmental conditions (Doelman and Breedveld (1999). For that reason in figure 1 the biodegradation of alpha-HCH under various conditions is given.

Already in 1973 Dennis Focht unravelled the degradation pathway of DDT, due to the successive different environmental conditions and consequently the contribution of different micro-organisms. So principally the mineralization of

DDT in soil may occur, however in some areas DDT is still a threat in food chains. The biodegradation of dioxine by halorespiration has been suggested in 2003 by Bunge and co-workers. However in 1995 Peter Adriaens already indicated in that direction by showing partly dechlorination of dioxins.

The halorespiring bacteria *Dehalococcoides* may play a crucial role in the bioremediation of soil and groundwater contaminated with VOCl's and POP's. *Dehalocococoides* has been discovered all over the world. Recently a Halococcoides-like bacterial strain has been isolated from HCB (hexa chloro benzene) contaminated sediment, enable to grow on HCB (Van Eekert, 2004, personal communication).

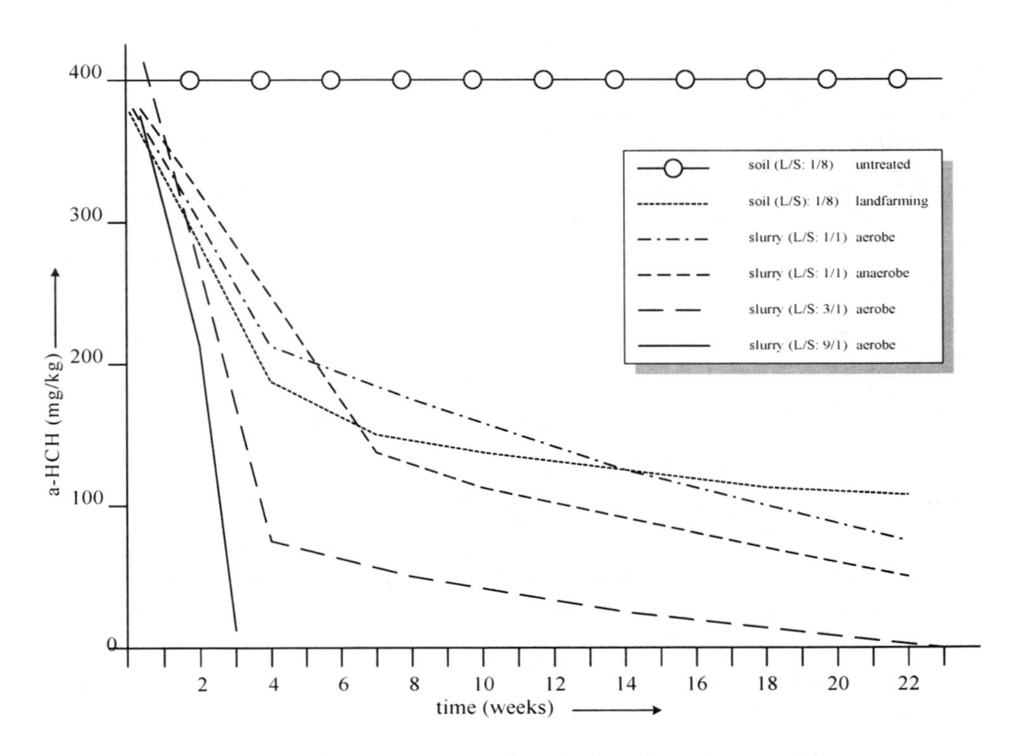

Figure 1. The biodegradation of alpha-HCH under various conditions

Since the biodegradation pathway of the various POP's can be very large and complex, it may be a suggestion to measure relevant intermediates, by checking the phase of degradation. Depending on environmental conditions chloro-phenoles and chloro-benzenes are considered to be key-intermediates. For several POP's the route to phenols has been given in figure 2.

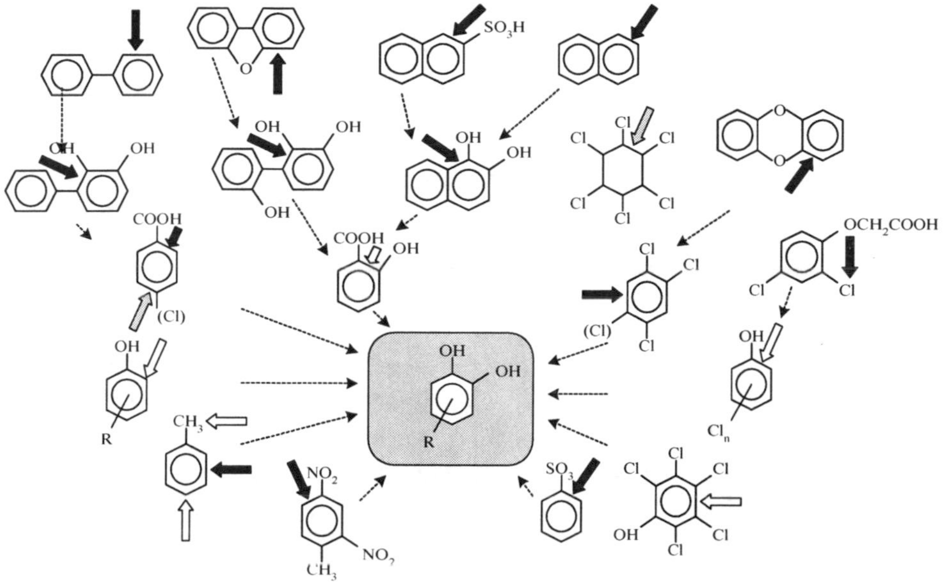

Figure 2. The degradation route of several POP's to phenoles (chlorinated)

Till the end of the 1970s the anaerobic microbial degradation of halogenated compounds such as PER (tetrachloroethene) and TRI (trichloroethene) was generally considered as "not occuring". In the eighties reductive dechlorination (Bouwer and McCarty, 1983) was discovered as a new phenomenon, leading to the formation of less harmful compounds. Then around 1992, also due to appearance of the PhD-theses of Holliger and others facts and principles on microbial degradation of chlorinated compounds became clearer.

The anaerobic microbial degradation of chlorinated xenobiotics became general knowledge, and applicable to bio-remediation. It became clear that bacteria can use PER (tetra chloro ethane) and other chlorinated compounds as electron acceptor. The role of the various prevailing electron acceptors, such as O_2, NO_3^-, Fe_3^+, SO_4^{2-} and CO_2 and various electron donors became clear. For example the fact that in the presence of nitrate complete dehalogenation seems to be impossible. The principle of "halorespiration" was added to the phenomenon of recycling elements. Suitable electron-donors and suitable environmental conditions became aspects to study to apply dechlorination in in-situ bioremediation processes. BioSoil R&D developed together with the laboratory of Microbiology of the Wageningen University a technique for the anaerobic dehalogenation of PER, TRI (tri chloro ethene) and a number of chlorinated hydrocarbons. This technique is now widely applied both in- and outside The Netherlands. The determination of the intermediates and their ratio show the degree of degradation. The degradation pathway is:

C = C ⟶ C = C ⟶ C = C ⟶ C = C ⟶ C = C

PER TRI DCE VC Ethene

Biodegradation Rates

From ten VOCl in-situ bioremediation locations in the city of Rotterdam (the Netherlands) Den Haan (2003) calculated the average biodegradation rate as first order constant. These values are given in table 3. The groundwater conditions were sulphate-reducing and methanogenic, while the DOC (dissolved organic content) was rather high due to the soil type, which was predominantly peat and loam.

TABLE 3. First order degradation rates of VOCl's

	Calculated rates from literature; until 1997	**Degradation rate in Rotterdam groundwater; until 2003**	**Degradation rate after introduction of Dehalococcoides (Bemmel & Klijn, 2006)**
PER	0.0023	0.0088	0.09
TRI	0.0033	0.0076	–
DCE	0.0038	0.0094	–
VC	0.0015	0.0049	–

The rates in Rotterdam are about 3 times higher than average literature value (Table 4). Those field rates are far lower than rates determined under laboratory conditions. Rates from laboratory data can not be translated to field expectations on an one to one basis. Nowadays (2006) bioremediation of PER can be called in The Netherlands "proven technology". This due to changing environmental conditions towards extreme anoxic conditions by addition of organinc matter the bacterial strain Dehalococcoides, as published by Bemmel and Klijn (2006). In principle the biodegradation of complex organics in the field (in-situ) is slow, as shown by Van der Werff (1992) for vegetation types.

TABLE 4. Some typical examples of degradation rate constants (k) of organic matter of defined vegetations (modified after Van der Werff, 1992)

Type of predominant vegetation	***k***	**Type of predominant vegetation**	***K***
Phragmites australis	0,0035	*Scirpus americanus*	0,0021-0,0025
Phragmites karka	0,0045	*Scirpus mucronatus*	0,0044
Typha domingensis	0,0078	*Juncus squarrosus*	0,0013
Typha glauca	0,0014	*Juncus roemerianus*	0,0016-0,0017
Typha Latifolia	0,0043	*Paspalum repens*	0,00717
Typha angustata	0,006	*Carex rostrata*	0,0046
Typha elephantina	0,0038	*Carex riparia*	0,0029
Scirpus fluviatilis	0,0018	*Zizania aquatica*	0,077

3. Fundamental Knowledge and Practical Monitoring Options

When a soil adapts to changes by microbial degradation of the contaminant, binding or becoming resistant it shows its vitality. Fatal may be sudden changes leading to another functioning (Ernst, 2002) or even to no functioning at all. It is the author's view that the state of the art on soil functioning is sufficient to predict the consequences of soil contamination. There is practical experience, all over the world, to predict where sustainable soil functioning is under threat due to contamination. Judging the sustainability, the health, the value or the functioning should not only contain that chemical aspects but also soil physical, soil chemical and soil biological ones (Figure 3).

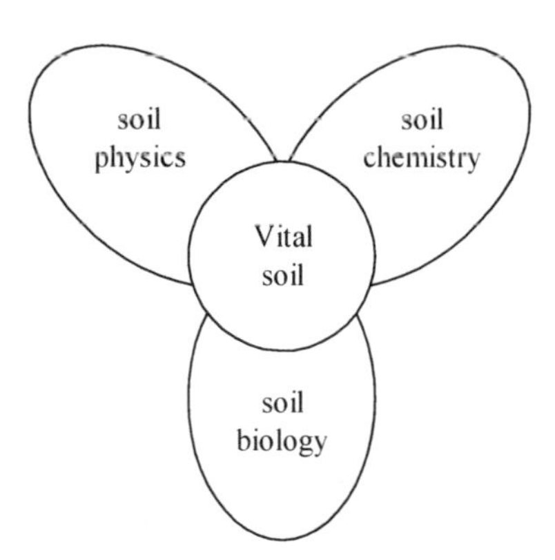

Figure 3. The golden triangle to monitor soil vitality

Implementation of bioremediation techniques demands those controlling monitoring options. Only in combination they are the basic elements of soil vitality. In relation to remediation the presence and action of the relevant microflora can be determined by molecular techniques. The basic soil chemical properties are soil type with carbon-content and -turn over, nitrogen and phosphorous content, and electron acceptors. Soil structure relates to physical properties. The bulk density of a soil is extremely important, but too often neglected in soil ecological contemplation and monitoring options. Waters and Oades (1991) show very nicely that the pore size of soils determines the presence possibilities of soil biota. This adds to the earlier emphasized saying that environmental conditions are the keys to specific aimed functioning of the soil system. The biological aspects contain microbial biomass, species diversity and functional diversity of earthworms and nematodes and many others. Specific monitoring sets can be attuned to various contamination groups (Table 5) (Doelman, 2004). For detailed information beyond this table 5, the various chapters Vital Soil (Doelman and Eijsackers, 2004) are recommended, since this table more or less combines in a synthesized way all chapters.

TABLE 5. Specific monitoring sets attuned to various contamination compound groups

<table>
<tr><td colspan="2">Carbon content
N content
Soil type
Soil compaction (structure)
Soil pH, redox
Landscape structure</td></tr>
<tr><td>Heavy metals</td><td>POPs (+ PAHs)</td></tr>
<tr><td>Chemical litter composition
Bacterial biomass
Bacteria: ratio sensitive/resistance
Nematodes: MI and functional diversity
Earthworms: species diversity and bio-accumulation
Soil fauna
Food web accumulation
Extractable level heavy metals</td><td>Bacteria: catabolic genes
Biodegradation rate
Earthworms: species diversity and bio-accumulation
Food web accumulation
Extractable level POP and PAH</td></tr>
</table>

Besides nematodes and earthworms the bio-accumulation of heavy metals and POP's and PAH's in food chains are illustrative mirrors of the quality of soil. The state of the art of nematodes overlaps the state of the practice, and has been applied all over the world (Bongers, 1990; Bongers and Ferris, 1999; Yeates and Bongers, 1999) Its application is based on species diversity and the functional meaning of the species composition (Figure 4).

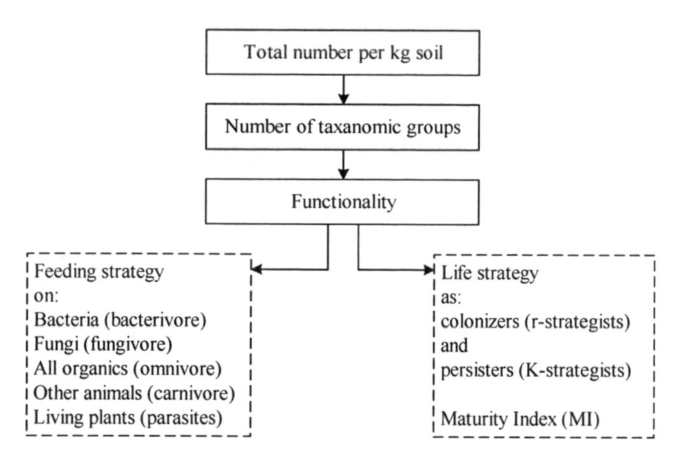

Figure 4. Nematode characterization on two aspects of functional diversity

The ambition for the soils system (what to plan, what to want, what to anticipate) determines the relevant monitoring schedule. The ambition to have or to maintain agriculture, is quite different from obtaining city green or obtaining an ecological system with an extensive biodiversity. Ecological recovery in general is a long lasting process of many years.

References

Adrians, P., et al., 1995. Bioavailability and transformation of highly chlorinated dibenzo-p-dioxins and dibenzofurans in anaerobic soils and sediments. Environmental Science and Technology 29: 2252-2260.

Alexander, M., 1985. Biodegradation of organic chemicals. Environmental Science and Technology 18: 106-117.

Atlas, R.M., 1981. Microbial degradation of petroleum hydrocarbons: an environmental perspective. Microbial reviews. 45: 180-209.

Bachmann, A., W. de Bruin, J.C. Jumelet, H.H.N. Rijnaarts and A.J.B. Zehnder, 1988. Aerobic biomineralization of alpha-hexachlorocyclohexanen in contaminated soil. Applied and Environmental Microbiology 54: 548-554.

Bemmel van, J.B.M. and R.B. Klijn, 2006. Fast biological remediation of chlorinated ethenes in Almelo. Bodem 6: 25-27.

Bongers, T., 1990. The maturity index: an ecological measure of environmental disturbance based on nematode species composition. Oecologia 83: 14-19.

Bongers, T. and H. Ferris, 1999. Nematode community structure as bio-indicator in environmental monitoring. Trends in Ecology and Evolution 14: 224-228.

Bouwer, E.J. and P.L. McCarty, 1993. Transformation of 1- and 2-carbon halogenated aliphatic compounds under methanogenic conditions. Applied and Environmental Microbiology 45: 1286-1294.

Bryson, B., 2003. A short history of nearly everything. Broadway Books New York ISBN 90 450 0970 6

Bunge, M.I., J.D. Kahn, E.K. Wallis and A.D. Wahner, 2003. Reductive halogenation of chlorinated dioxins by anaerobic bacterium. Nature 401.

De Duve, C., 1995. Vital Dust. The origin and evolution of life on earth. Life is a cosmic imperative. BasicBooks (Harper Collins Publishers, Inc.), New York.

Doelman, P. and G. Breedveld, 1999. In situ versus on site practises. In: Bioremediation of contaminated soils; Agronomy monograph 37: 539-558. Eds.: Adriano et al.; ASA, CSSA, SSSA, Madison Wisconsin, USA.

Doelman, P., 2004. Synthesis for soil management. In: Vital Soil: Developments in Soil Science Doelman & Eijsackers (eds). Elseviers Scientific Press, pp 313-338.

Doelman, P., E. Jansen, M. Michels and M. van Til, 1994. Effects of heavy metals in soil on microbial diversity and activity; the sensitivity/resistance index, an ecologically relevant parameter. Soil Biology and Soil Fertility 17: 177-184.

Doelman, P., L. Haanstra, H. Loonen and A. Vos, 1990. Decomposition of alpha- and beta-hexachlorocyclohexane in soil under field conditions in a temperate climate. Soil Biology and Biochemistry 22: 629-634.

Doelman, P. and H. Eijsackers. Vital Soil; function, value and properties. Developments in soil sciences, volume 29. Elseviers Scientific Press.

Eekert, van M.H.A., 2004. Personal communications.

Ernst, W.H.O., 1989. Mine vegetation in Europe. In: Shaw, A.J. (ed) Heavy metal tolerance in plants: Evolutionary aspects. CRC Press, Boca raton, pp 21-37.

Ernst, W.H.O., 2002. Living at the border of life. Retirement Oratio Vrije Universiteit, Amsterdam.

Haan, J. den., 2003. Natural Biodegradation of VOCl's in Rotterdam subsurface. Gemeentewerken Rotterdam report.

Holliger, H.C., 1992. Reductive dehalogenation by anaerobic bacteria. PhD thesis Wageningen University, the Netherlands.

Hong, S., J.P. Candelone, C.C. Patterson and C.F. Boutron, 1996. History of ancient copper smelting pollution during Roman and medieval times recorded in green land. Atmosphere and Environments 31: 2235-2242.

Hopkin, S.D.P., 1994. Effects of metal pollution on decomposition processes in terrestrial eco-systems with special reference to fungivorous soil arthropods. In: Metals in soil/plantsystems (ed. S.M. Ross) John Wileyand Sons, Chichester, pp 303-326.

Waters, A.G. and J.M. Oades, 1991. Organic matter in water-stable aggregates. In: Advances in Soil Organic Matter Research: The impact on agricculture and the environment. Wilson, W.S. (ed.). The Royal society of Chemistry, Cambridge, pp 113-174.

Werff, van der, P.A., 1992. Applied soil ecology in alternative agriculture. Reader University Wageningen, the Netherlands.

Wilson, J.T., L.E. Leach, M. Henson and J.N, Jones, 1986. In situ biorestoration as a groundwater remediation technique. Groundwater Monitoring Review 6: 56-64.

Yeates, G.W. and T. Bongers, 1999. Nematode diversity in agro-ecosystems. Agriculture, Ecosystem and Environment 74: 113-135.

CHEMICAL OXIDATION FOR CLEAN UP OF CONTAMINATED GROUND WATER

ROBERT L. SIEGRIST[1*], MICHELLE L. CRIMI[2], JUNKO MUNAKATA-MARR[1], TISSA ILLANGASEKARE[1], PAMELA DUGAN[1], JEFF HEIDERSCHEIDT[3], BEN PETRI[1] AND JASON SAHL[1]

[1]*Environmental Science and Engineering Division, Colorado School of Mines, Golden, CO, 80401 USA*

[2]*Department of Environmental Health, East Tennessee State University, Kingsport, TN 37663 USA*

[3]*Air Force Institute of Technology, Wright-Patterson Air Force Base, OH 45433 USA*

Abstract. Contamination of soil and ground water by organic chemicals is a widespread problem across the U.S. and around the world. At sites where organic chemicals are present in the form of dense nonaqueous phase liquids (DNAPLs), clean up of contaminated ground water has been extremely difficult and costly; conventional ground water pumping and treatment approaches have commonly failed to achieve clean up goals. Major research and development efforts have been directed at finding alternative remedies that can clean up ground water and eliminate risks or reduce them to an acceptable level. Recent efforts have increasingly focused on source zone treatment of DNAPL contamination to reduce the volume and mass of DNAPLs available for dissolution into ground water. A variety of in situ technologies have been developed and demonstrated, including in situ chemical oxidation (ISCO). ISCO involves the delivery of chemical oxidants into the subsurface to destroy organic chemicals (e.g., chlorinated organic solvents or fuels) and thereby remediate a site to a risk-based clean up goal. ISCO can feasibly be implemented alone to

[*] To whom correspondence should be addressed. Robert L. Siegrist, Professor and Director, Environmental Science and Engineering, Colorado School of Mines, 206 Coolbaugh Hall, Golden, CO 80401-1887, USA; e-mail: siegrist@mines.edu

M.D. Annable et al. (eds.), Methods and Techniques for Cleaning-up Contaminated Sites, 45–58.

treat a contaminant source zone or an associated groundwater plume, or used in combination with other remedial technologies (e.g., after surfactant flushing or before bioremediation).

Keywords: Remediation; risk reduction; trichloroethene; tetrachloroethene; Colorado School of Mines

1. Ground Water Contamination by Organic Chemicals

Contamination of soil and ground water by organic chemicals represents a major environmental problem for many sites throughout the United States and in other industrialized nations (Kavanaugh et al. 2003, GAO 2005). In the U.S. today, there are still an estimated 30,000 to 50,000 sites with ground water contamination (excluding petroleum contamination from underground storage tanks) (Kavanaugh et al. 2003). About 80% of the sites are contaminated with organic chemicals, and of these, 60% likely have dense nonaqueous phase liquids (DNAPLs) present. Common DNAPLs include trichloroethene (TCE), tetrachloroethene (PCE), chloroform, and carbon tetrachloride. DNAPLs present a long-term problem due to their chemical properties, including low solubility and high density, which can contribute to high contaminant concentrations in ground water over extensive time frames. Contamination of subsurface soils and ground water can present unacceptable risks to human health and environmental quality (e.g., increased cancer risk through ingestion of contaminated drinking water or inhalation of vapors) (Figure 1).

An early approach to clean up of ground water contaminated by organic chemicals involved using pumping wells for extracting ground water and treating it above-ground. But experiences with this approach revealed it to have severe limitations and high costs (Mackay and Cherry 1989, USEPA 1999). The cleanup costs for 15,000 to 25,000 sites with DNAPLs are known to be significant when conventional ground water pumping and treatment systems are employed. The median cost to operate a ground water pump and treat system is on the order of $180,000 per year with a range of $30,000 to $4,000,000. For all sites combined, the annual costs are $2.7 to 4.5 billion dollars per year. Assuming a 30-year life and a 5 to 10% interest rate, life-cycle costs of "cleanup" could range from $50 to $100 billion dollars. Many now conclude that clean up based on ground water pump and treat alone is hopeless and alternative technologies and approaches are needed (Kavanaugh et al. 2003).

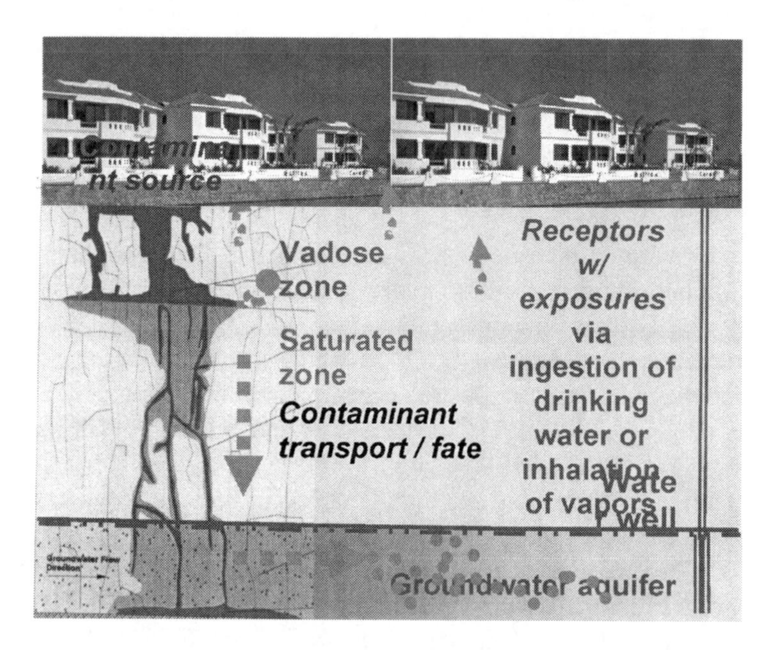

Figure 1. Illustration of a DNAPL source zone leading to long-term contamination of subsurface soil and ground water and resultant risks to public health.

Major research and development efforts have been directed at finding alternative remedies that can clean up ground water and eliminate risks or reduce them to an acceptable level (NRC 1994, 1997). Recent efforts have increasingly focused on source zone treatment of DNAPL contamination to reduce the volume and mass of DNAPLs available for dissolution into ground water. A variety of in situ technologies have been developed and demonstrated and, to varying degrees, utilized for cleanup of DNAPL contaminated sites, including enhanced recovery using in situ thermal or surfactant/cosolvent flushing methods or in situ destruction using chemical or biological methods.

2. General Description of Chemical Oxidation as an In Situ Remediation Technology

A wide range of organic chemicals are susceptible to oxidative degradation through chemical reactions with oxidants such as catalyzed H_2O_2 (also known as modified Fenton's reagent), potassium and sodium permanganate ($KMnO_4$, $NaMnO_4$), sodium persulfate ($Na_2S_2O_8$), and ozone (O_3). Under the right conditions, these oxidants can mineralize many contaminants including chlorinated hydrocarbons (e.g., TCE, PCE), fuels (e.g., benzene, toluene, MTBE), phenols (e.g., pentachlorophenol), PAHs (e.g., naphthalene, phenanthrene, PCBs), explosives (e.g., TNT), and pesticides (e.g., lindane). Based on extensive laboratory experiments, reaction stoichiometries, pathways, and kinetics have been established for many common chemicals of concern. Degradation reactions tend to involve electron transfer or free radical processes, involve simple to complex pathways with various intermediates, and follow 2nd order kinetics. The need for activation and the sensitivity to matrix conditions such as temperature, pH, and salinity vary between the different oxidants.

In situ chemical oxidation (ISCO) involves the introduction of chemical oxidants into the subsurface to destroy organic contaminants in soil and ground water (e.g., USEPA 1998, ESTCP 1999, Yin and Allen 1999, Siegrist et al. 2000, 2001, ITRC 2001, 2005) (Figure 2). The ISCO systems that can be (and have been) applied in the field are highly varied in their features based on site conditions and cleanup goals. Clean up goals tend to be site specific, but they generally fall into three major categories: (1) reduce the contaminant concentration or mass in the ISCO-treated zone by some % (e.g., >90%); (2) achieve a specified post-ISCO contaminant concentration in ISCO-treated soil (e.g., <1 mg/kg) or ground water (e.g., <100 ug/L), or (3) achieve a concentration in a ground water plume at some compliance plane down gradient from an ISCO-treated DNAPL source zone. Oxidants can be used alone or in combination and can be delivered into the subsurface at widely varied concentrations and mass flow rates in liquid, gaseous and solid phases through different subsurface delivery involving passive or active (one or more) injection campaigns (Siegrist

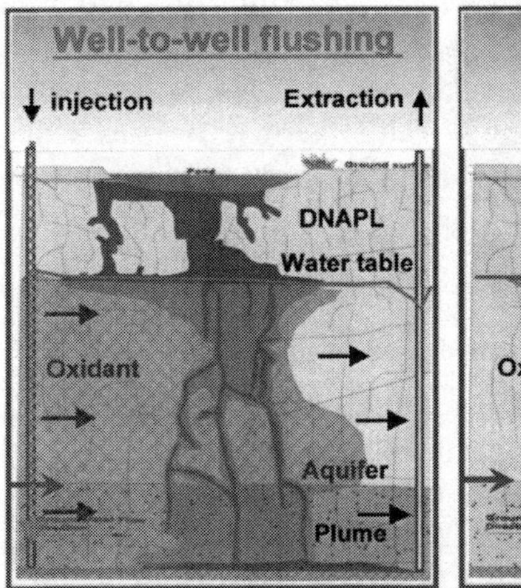

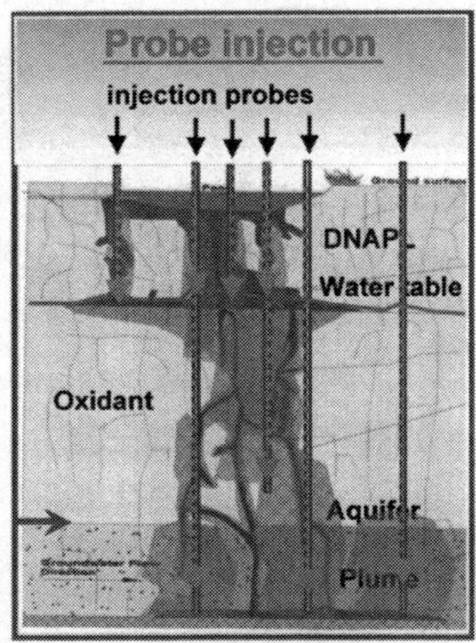

Figure 2. Illustration of in situ chemical oxidation where well-to-well flushing or direct-push probes are used to deliver oxidants into the subsurface to clean up soil and groundwater contamination.

et al. 2001, ITRC 2005). Methods of oxidant delivery have included permeation by vertical injection probes, deep soil mixing, flushing by vertical and horizontal ground water wells, and reactive zone emplacement by hydraulic fracturing.

During the past 10 years, field applications of ISCO have increased across the U.S. and abroad. At some sites, ISCO has been applied and the destruction of contaminants of concern (COCs) has occurred such that clean up goals have been met in a cost-effective and timely manner. However, at other sites, ISCO applications have had uncertain or poor in situ treatment performance. Poor performance has often been attributed to poor uniformity of oxidant delivery caused by low permeability zones and site heterogeneity, excessive oxidant consumption by natural subsurface materials, or presence of large masses of dense nonaqueous phase liquids (Siegrist et al. 2001, Siegrist et al. 2006). DNAPLs can exhibit unstable behavior resulting in complex entrapment architecture within source zones (Held and Illangasekare 1995, Illangasekare et al. 1995). The aquifer heterogeneity and the DNAPL entrapment itself can affect the flow field that determines the rate of DNAPL dissolution and delivery of treatment agents (Saba and Illangasekare 2000, Saenton et al. 2001). In some applications there have been concerns over secondary effects such as mobilization of metals, loss of well-screen and formation permeability, and gas evolution and fugitive emissions, as well as health and safety practices (Siegrist et al. 2001, Crimi and Siegrist 2003).

ISCO system selection, design, and implementation practices must rely on a clear and thorough understanding of ISCO and its applicability to a given set of contaminant and site conditions to achieve site-specific cleanup goals. A number of key issues may be relevant and need to be addressed during the selection, design, and implementation of ISCO, regardless of the oxidant and delivery system being employed, including (1) amenability of the target contaminants to oxidative degradation, (2) effectiveness of nonaqueous phase liquid (NAPL) destruction, (3) optimal oxidant loading (dose concentration and delivery) for a given organic in a given subsurface setting, (4) unproductive oxidant consumption, (5) potential adverse effects (e.g., mobilizing metals, forming toxic byproducts, reducing formation permeability, generating off-gases and heat), and (6) potential for ISCO to be coupled with other remediation technologies. Matching the oxidant and delivery system to the COCs and site conditions is critical to achieve site-specific performance goals.

3. ISCO Research at the Colorado School of Mines

A major program of research and development concerning characterization and remediation of contaminated soil and ground water has been ongoing at the Colorado School of Mines (CSM) for more than 10 years. One of the major research thrusts has focused on advancing the science and engineering of ISCO, utilized alone or in combination with other remedial approaches like bio-remediation (e.g., Crimi and Siegrist 2003, 2004a,b, 2005, Lowe et al. 2002, Ross et al. 2005, Sahl and Munakata-Marr 2006, Siegrist et al. 1999, 2000, 2002, Urynowicz and Siegrist 2005). Within this thrust area, a large research project was recently completed for the U.S. Strategic Environmental Research and Development Program (SERDP) (see www.serdp.org): "Reaction and Transport Processes Controlling In Situ Chemical Oxidation of DNAPLs" (SERDP project CU-1290) (Siegrist et al. 2006). SERDP project CU-1290 encompassed an extensive array of experimental work and modeling efforts to enhance the understanding of the pore/interfacial scale DNAPL reactions and porous media transport processes that govern the delivery of oxidant to a DNAPL-water interface and the degradation of the DNAPL present in the subsurface at a contaminated site. These efforts examined the application of potassium permanganate (PP) or catalyzed hydrogen peroxide (CHP) for remediation of PCE or TCE DNAPL contamination. A summary of the methods and results of the research completed during this CSM project is given below; details regarding the methods and results of various elements of the research are presented elsewhere in existing or forthcoming publications (Table 1).

3.1. CSM RESEARCH APPROACH

During SERDP project CU-1290, experimental work and modeling studies were completed at CSM using an array of apparatus and techniques (Figure 3). The fundamental research completed during this project was designed to provide a knowledge base for the future development of guidance on the principles and practices of ISCO so that it can be selected as a preferred remedy when appropriate, either as a stand-alone method or combined with other remediation technologies. The scope of research activities completed during SERDP project CU-1290 included the following:

- Bench-scale studies involving aqueous phase vial reactors (VRs) were used to determine how dissolved phase reaction kinetics for PCE and TCE are affected by the composition of the bulk aqueous phase (i.e., oxidant type and concentration and natural organic matter (NOM), mineralogy, pH). Additional studies conducted in multiphase vial reactors (MVRs) investigated the impact of ISCO on TCE and PCE DNAPL mass transfer rates as a function of aqueous phase compositions.
- Bench-scale studies involving slurry reactors and 1-D diffusion cells were used to determine the effects of porous media of varying properties on oxidant transport and degradation of DNAPLs. These were conducted under

contrasting transport regimes of either completely mixed systems or static, diffusion-controlled systems. Zero-headspace reactors (ZHRs) were utilized to investigate the impact of differing conditions (e.g., PCE vs. TCE at different mass levels, presence and composition of NOM, oxidant type and loading) on the efficiency and effectiveness of ISCO within well-mixed soil slurry systems. Steady-state diffusion tests were performed to determine the viability of diffusive transport of oxidants for contaminant destruction in low permeability media. Impacts to porous media properties due to of oxidation were also determined. As part of these studies, it was necessary to evaluate the oxidant persistence in porous media (i.e., loss due to natural oxidant demand (NOD) or oxidant decomposition) and determine if CHP could be readily stabilized to promote subsurface transport.

- Bench and intermediate scale flow-through studies were used to evaluate the reaction and transport processes that occur at the pilot-scale, including the effects of ISCO systems of varied designs on different DNAPL mass and entrapment architectures. Bench scale 1-D flow-through tube reactor (FTR) studies were utilized to investigate the impact of varied oxidant delivery velocities and concentrations on DNAPL mass depletion rates and efficiencies in systems with varied DNAPL types and entrapment architectures. Further experimentation was conducted in 2-D flow cells, which investigated varied oxidant types and delivery techniques on an experimental system with a more complex source architecture and flow field.
- Several intermediate-scale and large-scale 2-D tank systems were used to investigate complex DNAPL entrapment architectures and complex heterogeneous flow fields to determine impacts to DNAPL mass depletion and mass flux resultant of ISCO using permanganate, including the potential reduction in permeability due to manganese dioxide deposition. These experiments enabled the testing of a new model, which was developed as part of this project, as well as in cooperation with SERDP project CU-1294: "*Mass Transfer from Entrapped DNAPL Sources Undergoing Remediation: Characterization Methods and Prediction Tools.*" This CSM model is named CORT3D, and it is based on the RT3D modeling code. CORT3D was developed to account for permanganate oxidation chemistry, multiple rate-limited NOD fractions, DNAPL mass transfer, and changes in porous media flow properties resulting from PP oxidation.
- Batch and flow-through experiments were completed to evaluate the viability and effectiveness of coupling ISCO with other remediation technologies. Studies were conducted to investigate the feasibility and design considerations of coupling ISCO with both anaerobic and aerobic bioremediation technologies, utilizing batch microcosm studies and 1-D

column experiments. Experimentation also investigated coupling ISCO with surfactant and cosolvent remediation. Batch screening studies were conducted to determine which surfactants and cosolvents were most compatible with the oxidants in this study, and follow up experimentation was conducted within a 2-D flow cell to determine the effectiveness of various coupling regimes (i.e., sequential versus combined delivery of surfactants/ cosolvents and oxidants).

- A final aspect of the project included experiments designed to evaluate whether partitioning tracer test methods provide a viable way to measure DNAPL mass removals after ISCO has been applied to a DNAPL site. Experiments were conducted to determine how residual oxidant, including MnO_2 deposition at the NAPL-water interface, might impact alcohol and ketone tracers and their respective NAPL-water partitioning coefficients. Effective tracers were then applied in 2-D flow cell systems to determine how effective they are at measuring performance at ISCO treated sites.

TABLE 1. Summary of research elements completed at CSM during SERDP project CU-1290 and current sources of information that describe the relevant methods and results.

CSM research element within project CU-1290	CSM student M.S. or Ph.D. thesis and journal papers and selected conference papers (as of July 2006)
Bench scale kinetic and DNAPL degradation studies: vial reactors and multiphase vial reactors	Jackson 2004, Crimi and Siegrist 2005
Bench-scale studies of porous media effects on oxidation reactions: zero headspace reactors, 1-D diffusion cells	Seitz 2004, Crimi and Siegrist 2005
Upscaling reaction/transport experiments: flow-through tube reactors, 2-D flow-through cells, 2-D tanks	Heiderscheidt 2005, Petri 2006
Experimental evaluation of coupling ISCO with post-ISCO bioremediation: batch microcosms, 1-D columns	Sahl 2005, Sahl and Munakata-Marr 2006, Sahl et al. 2006
Experimental evaluation of coupling ISCO with pre-ISCO surfactant/ cosolvent flushing: batch vial reactors, 2-D flow-through cells	Dugan 2006
Experimental studies of the compatibility of partitioning tracers and ISCO: batch vial reactors, 2-D flow-through cells	Dugan 2006
Mathematical modeling	Heiderscheidt 2005, Petri 2006

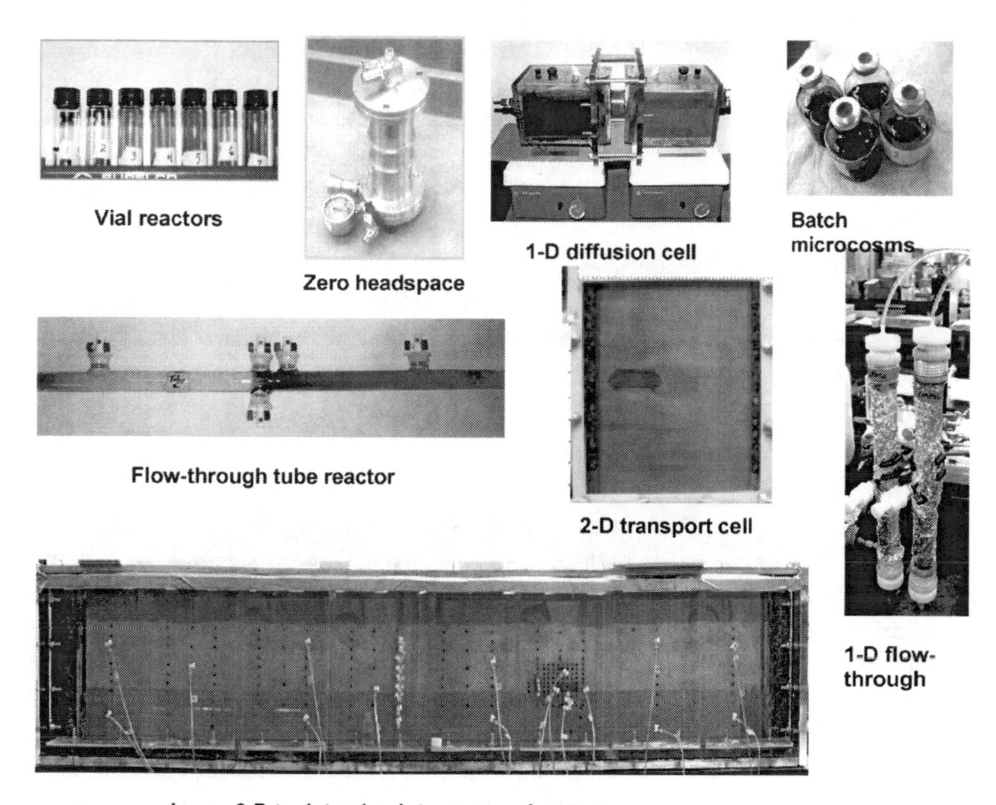

Figure 3. Illustration of experimental apparatus used during CSM research to examine ISCO as a remediation technology to clean up organic chemicals in soil and ground water.

3.2. SUMMARY OF CSM FINDINGS

The major findings of research completed at CSM during SERDP project CU-1290 encompass the rate and extent of PCE and TCE DNAPL degradation as affected by oxidant type and concentration, delivery method, and subsurface conditions. Findings also address the viability of coupling ISCO with other remediation technologies. Listed below are several generalized conclusions concerning the application of ISCO to DNAPL sites based on the research completed at CSM.

- CHP was found to have very fast rates of reaction, and subsequently fast rates of oxidant decomposition. As such, the oxidant persistence as evaluated in these studies was poor. This was found to hold true in all systems of varying experimental scale that evaluated the use of CHP. In general, systems that provided effective contact of CHP with the contaminant appeared to be highly efficient with regard to contaminant destruction and oxidant use (i.e. RTE values in the VRs, MVRs and ZHRs). However, the effectiveness (rate and extent) of contaminant mass depletion of CHP was generally lower than in equivalent PP systems in every experimental series CHP was evaluated in (VRs, MVRs, ZHRs, diffusion cells, 2-D flow cells). The lower effectiveness of CHP is largely due to difficulties that arose in optimizing oxidant delivery and oxidant activation.

- PP, in contrast to CHP, was much more effective in reducing contaminant mass in the experimental systems evaluated. Due to the reduced reactivity of permanganate, advective and diffusive transport of oxidant is possible, as evidenced in all experimental transport studies (e.g., 1-D diffusion cells, FTRs, 2-D cells).
- When considering remediation effectiveness, oxidant concentration can interact strongly with the velocity of oxidant delivery into a porous media system. For either oxidant studied in this project, in porous media systems where delivery to the DNAPL depended on advective transport of the oxidant (e.g., FTRs or 2D cells), the use of lower concentrations of oxidant delivered at higher fluid velocities appeared preferable. A common theme among the experiments that evaluated multiple velocities was that higher velocities led to more effective remediation performance, evidenced by increased DNAPL mass depletion, and frequently reduced adverse impacts and enhanced potential for bio-coupling. However, higher oxidant concentrations can promote diffusive transport of oxidant into low permeability media and be more time-efficient from a site flushing perspective as fewer pore volumes may be required to destroy a given DNAPL contaminant source.
- Natural oxidant demand (for PP) and oxidant decomposition (for CHP) increased with increasing complexity of porous media (e.g., higher reduced mineral content, NOM, or clay/silt particle fractions), increasing oxidant concentration, and increased contact time. NOD and oxidant decomposition challenge both advective and diffusive oxidant transport. However, site-specific characterization in a way that allows for prediction of transport in the subsurface is difficult. Conditions used during lab testing of oxidant-subsurface interactions need to mirror the planned oxidant delivery and transport in the subsurface.
- TCE was found to be depleted and destroyed faster than PCE in every experimental system (VRs, MVRs, ZHRs, diffusion cells and FTRs) that evaluated both DNAPL types. This is likely due to the order of magnitude higher solubility of TCE versus PCE, as well the faster kinetic rate of reaction of TCE compared to PCE in permanganate systems. However, the

solubility limit can have secondary impacts on ISCO system performance. Based on the FTR experiments, TCE was more likely than PCE to result in CO_2 gas generation from permanganate ISCO. This is largely due to the fact that higher contaminant concentrations in TCE systems result in more concentrated proton generation, as well as higher concentrations of aqueous CO_2. The resulting pH change, if the system is not adequately buffered, can result in off-gassing, which may mobilize DNAPL or vapor phase contaminant, as well as reduce hydraulic conductivity. However, such effects may be compound-specific. PP oxidation of *cis*-DCE generates alkalinity instead of acidity and would be less likely to cause this problem, despite the higher solubility limit of *cis*-DCE.

- DNAPL present in the subsurface as residual ganglia are highly amenable to depletion and destruction by PP flushing. However, when pooled DNAPL was present, oxidation effectiveness was more variable. Residuals are depleted much faster than pools due to their high surface area to volume ratio. Mass transfer was observed to be enhanced from residuals by PP oxidation by bulk aqueous phase contaminant oxidation, and possibly other impacts to mass transfer processes, based on FTR and CORT3D modeling study results.
- Natural organic matter and certain minerals (i.e., goethite) present in the subsurface can complicate the understanding of ISCO efficiency and effectiveness. For example, the results of this study revealed that NOM in porous media during ISCO has a more complex effect on remediation performance than simply increasing natural oxidant demand.
- Coupling ISCO with anaerobic reductive dechlorination for destruction of contaminants such as TCE and PCE is a viable approach. In general, CHP oxidation is disruptive to anaerobic microbes due to oxygenation of the subsurface, but chemical oxidation of porous media by CHP can generate higher levels of dissolved organic carbon (DOC), which can provide a substrate for anaerobic microbes growing in down-gradient areas where conditions remain anaerobic. Permanganate oxidation appears to be more amenable to coupling with anaerobic reductive dechlorination processes. For an anaerobic mixed culture, reductive dechlorination behavior ceased after oxidant flushing began but rebounded after oxidation ceased. The amount of time until activity rebound varied between systems and appeared to depend on the total time and concentration of permanganate to which the microbes were exposed.
- Coupling ISCO with surfactant enhanced aquifer restoration and use of PTT test methods appears viable, but only with careful evaluation and design to ensure that the oxidant-surfactant and oxidant-PTT systems are compatible.

The fundamental research completed at CSM has provided a knowledge base for the future development of guidance on the principles and practices of ISCO so that it can be selected as a preferred remedy when appropriate and can be implemented to reliably achieve performance goals. Guidance is needed, including decision aids and design tools, to enable cost-effective implementation of ISCO for clean up of ground water at a given site, using ISCO either as a stand-alone method or by coupling it with a pre- or post-ISCO operation. This needed guidance is being developed by the CSM project team under a follow-on project: "*In Situ Chemical Oxidation for Remediation of Groundwater: Technology Practices Manual*" with funding from the U.S. Environmental Security Technology Certification Program (ESTCP) (see www.estcp.org).

References

Crimi, M.L. and Siegrist, R.L. 2003. Geochemical effects associated with permanganate oxidation of DNAPLs. Ground Water. 41(4):458-469.

Crimi, M.L. and Siegrist, R.L. 2004a. Association of cadmium with MnO2 particles generated during permanganate oxidation. Water Research. 38(4):887-894.

Crimi, M.L. and Siegrist, R.L. 2004b. Impact of reaction conditions on MnO2 genesis during permanganate oxidation. Journal Environmental Engineering. 130(5):562-572.

Crimi, M.L. and Siegrist, R.L. 2005. Factors affecting effectiveness and efficiency of DNAPL destruction using potassium permanganate and catalyzed hydrogen peroxide. J. Environmental Engineering, 131(12):1724-1732.

Dugan, P. 2006. Coupling in situ technologies for DNAPL remediation and viability of the PITT for post-remediation performance assessment. Ph.D. dissertation, Environmental Science and Engineering Division, Colorado School of Mines, Golden, CO. August 2006.

Environmental Security Technology Certification Program (ESTCP). 1999. Technology status review: In situ oxidation. http://www.estcp.gov.

GAO. 2005. United States Government Accountability Office (GAO) Report to Congressional Committees. Groundwater contamination: DOD uses and develops a range of remediation technologies to clean up military sites. GAO-55-666 Groundwater Contamination.

Gates-Anderson, D.D., Siegrist, R.L., and Cline, S.R. 2001. Comparison of potassium permanganate and hydrogen peroxide as chemical oxidants for organically contaminated soils. J. Environmental Eng. 127(4):337-347.

Heiderscheidt, J.L. 2005. DNAPL source zone depletion during in situ chemical oxidation (ISCO): experimental and modeling studies. Ph.D. dissertation, Environmental Science and Engineering Division, Colorado School of Mines, Golden, CO. August 2005.

Held, R.J. and Illangasekare, T.H. 1995. Fingering of dense non-aqueous phase liquids in porous media 1. Experimental investigation. Water Resources Resh., 31(5):1213-1222.

Illangasekare, T.H., Ramsey, J.L., Jensen, K.H. and Butts, M. 1995. Experimental study of movement and distribution of dense organic contaminants in heterogeneous aquifers, J. Contaminant Hydrology, 20, 1-25.

Interstate Technology & Regulatory Council (ITRC) 2001. Technical and regulatory guidance for in situ chemical oxidation of contaminated soil and groundwater (ISCO-1). The Interstate Technology & Regulatory Cooperation Work Group In Situ Chemical Oxidation Work Team. www.itrcweb.org/gd_ISCO.asp.

Interstate Technology & Regulatory Council (ITRC) 2005. Technical and regulatory guidance for in situ chemical oxidation of contaminated soil and groundwater, 2nd Edition (ISCO-2). The Interstate Technology & Regulatory Council In Situ Chemical Oxidation Team. www.itrcweb.org/gd_ISCO.asp.

Jackson, S.F. 2004. Comparative evaluation of potassium permanganate and catalyzed hydrogen peroxide during in situ chemical oxidation of DNAPLs. M.S. Thesis, Environmental Science and Engineering Division, Colorado School of Mines, Golden, CO. January 2004.

Kavanaugh, M.C., Rao, P.S.C., Abriola, L., Cherry, J., Destouni, G., Falta, R., Major, D., Mercer, J., Newell, C., Sale, T., Shoemaker, S., Siegrist, R.L., Teutsch, G. and Udell, K. 2003. The DNAPL cleanup challenge: source removal or long term management. Report of an Expert Panel to the U.S. EPA National Risk Management Laboratory and Technology Innovation Office. EPA/600/R-03/143, December 2003.

Lowe, K.S., Gardner, F.G. and Siegrist, R.L. 2002. Field pilot test of in situ chemical oxidation through recirculation using vertical wells. J. Ground Water Monitoring and Remediation. Winter issue. pp. 106-115.

MacKay, D.M. and Cherry, J.A. 1989. Ground water contamination: limits of pump-and-treat remediation. Environ. Sci. Technol. 23:630-636.

National Research Council (NRC). 1994. Alternatives for ground water cleanup. National Academy Press, Washington, D.C.

National Research Council (NRC). 1997. Innovations in ground water and soil cleanup. National Academy Press, Washington, D.C.

Petri, B.G. 2006. Impacts of subsurface permanganate delivery parameters on dense nonaqueous phase liquid mass depletion rates. M.S. thesis, Environmental Science and Engineering Division, Colorado School of Mines, Golden, CO. January 2006.

Ross, C., Murdoch, L.C., Freedman, D.L. and Siegrist, R.L. 2005. Characteristics of potassium permanganate encapsulated in polymer. J. Environmental Engineering. 131(8):1203-1211.

Saba. T and Illangasekare, T.H. 2000. Effect of groundwater flow dimensionality on mass transfer from entrapped nonaqueous phase liquids. Water Res. Resh., 36(4):971-979.

Saenton, S., Illangasekare, T.H., Soga, K., and Saba, T.A. 2001. Effects of source zone heterogeneity on surfactant enhanced NAPL dissolution and resulting remediation end-points, J. of Contaminant Hydrology. 59:27-44.

Sahl, J. 2005. Coupling in situ chemical oxidation (ISCO) with bioremediation processes in the treatment of dense non-aqueous phase liquids (DNAPLs). M.S. thesis, Environmental Science and Engineering Division, Colorado School of Mines, Golden, CO. April 2005.

Sahl, J., and Munakata-Marr, J. 2006. The effects of in situ chemical oxidation on microbial processes: A Review. Remediation Journal. 16(3):57-70.

Sahl, J.W., Munakata-Marr, J., Crimi, M.L. and Siegrist, R.L. 2006. Coupling Permanganate Oxidation with Microbial Dechlorination of Tetrachloroethene. Water Environment Research. Accepted and in press.

Schnarr, M.J., Truax, C.L., Farquhar, G.J., Hood, E.D., Gonullu, T. and Stickney, B. 1998. Laboratory and controlled field experiments using potassium permanganate to remediate trichloroethylene and perchloroethylene DNAPLs in porous media. J. Contam. Hydrol. 29(3):205-224.

Seitz, S.J. 2004. Experimental evaluation of mass transfer and matrix interactions during in situ chemical oxidation relying on diffusive transport. M.S. thesis, Environmental Science and Engineering Division, Colorado School of Mines, Golden, CO. December 2004.

Siegrist, R.L., Lowe, K.S., Smuin, D.R., West, O.R., Gunderson, J.S., Korte, N.E., Pickering, D.A. and Houk, T.C. 1998. Permeation dispersal of reactive fluids for in situ remediation: field studies. ORNL/TM-13596. Project Report prepared by Oak Ridge National Laboratory for the U.S. DOE.

Siegrist, R.L., Lowe, K.S., Murdoch, L.C., Case, T.L. and Pickering, D.A. 1999. In situ oxidation by fracture emplaced reactive solids. J. Environmental Engineering, 125(5):429-440.

Siegrist R.L., Urynowicz, M.A. and West, O.R. 2000. In situ chemical oxidation for remediation of contaminated soil and ground water. Ground Water Currents. Issue No. 37, US EPA Office of Solid Waste and Emergency Response, EPA 542-N-00-006. September, 2000. http://www.epa.gov/tio.

Siegrist R.L., Urynowicz, M.A., West, O.R., Crimi, M.L. and Lowe, K.S. 2001. Principles and Practices of In Situ Chemical Oxidation Using Permanganate. Battelle Press, Columbus Ohio. 336 pages.

Siegrist, R.L., Urynowicz, M.A., Crimi, M.L. and Lowe, K.S. 2002. Genesis and effects of particles produced during in situ chemical oxidation using permanganate. J. Environmental Engineering, 128(11):1068:1079.

Siegrist, R.L., Crimi, M.L., Munakata-Marr, J., Illangasekare, T., Lowe, K.S., Van Cuyk, S., Dugan, P., Heiderscheidt, J., Jackson, S., Petri, B. Sahl, J. and Seitz, S. 2006. Reaction and transport processes controlling in situ chemical oxidation of DNAPLs. Final project report to the U.S. Strategic Environmental Research and Development Program (SERDP) for SERDP project CU-1290.

USEPA. 1998. In situ remediation technology: in situ chemical oxidation. EPA 542-R-98-008. Office of Solid Waste and Emergency Response. Washington, D.C.

USEPA. 1999. Ground water cleanup: overview of operating experience at 28 sites. EPA 542-R-99-006. Office of Solid Waste and Emergency Response. Washington, D.C.

Urynowicz, M.A. and Siegrist, R.L. 2005. Interphase mass transfer during chemical oxidation of TCE DNAPL in an aqueous system. J. Contaminant Hydrology. 80(3-4):93-106.

Yin, Y. and Allen, H.E. 1999. In situ chemical treatment. Ground Water Remediation Technology Analysis Center, Technology Evaluation Report, TE-99-01. July, 1999.

DETECTION AND MEASUREMENT TECHNIQUES TO IDENTIFY THE PRESENCE OF NAPLS IN THE FIELD

ILSE VAN KEER[1*], JAN BRONDERS[1], KAAT TOUCHANT[1], JEROEN VERHACK[1,2] AND DANNY WILCZEK[1]
[1] *VITO (Flemish Institute for Technological Research), Boeretang 200, B-2400 Mol, Belgium*
[2] *Catholic University Leuven, Celestijnenlaan 200E, B-3001 Heverlee, Belgium*

Abstract. The investigation of sites polluted with NAPLs (non aqueous phase liquids) such as BTEX (benzene, toluene, ethyl benzene and xylene) and chlorinated aliphatic hydrocarbons (CAH) is a real challenge. It has been demonstrated that the use of so called "conventional" field measurement techniques (drilling, placement of observation wells, ex-situ analysis of soil-groundwater samples) does not guarantee that a NAPL pollution is correctly characterized. To achieve a better detection it is advised to use conventional techniques in combination with "alternative" field measurement techniques such as MIP (Membrane Interphase Probe, SPGS (screen-point groundwater sampler) or CSIA (component-specific isotope analysis). The usefulness of these methods on the field is demonstrated for a selected site.

Keywords: LNAPL; DNAPL; field measurement techniques, screen-point groundwater sampler; Membrane Interphase Probe™; component-specific stable isotopes

1. Introduction

Field experience and several studies (e.g. Huling and Weaver, 1991 and Newell et al., 1995) demonstrate that when released into the environment, NAPLs migrate under the force of gravity and move through the unsaturated zone. A fraction will be retained by capillary forces as residual product in the soil

[*] To whom correspondence should be adressed. Ilse Van Keer, VITO (Flemish Institute for Technological Research), Boeretang 200, B-2400 Mol; ilse.vankeer@vito.be

M.D. Annable et al. (eds.), Methods and Techniques for Cleaning-up Contaminated Sites, 59–69.

pores. The capillary forces and the differences in grain size of the soil, also induce lateral spreading. If sufficient NAPL is released, it will migrate until it encounters for instance the water table or low permeable strata.

Once the capillary fringe is reached, light non-aqueous phase liquids (LNAPL) may move laterally as a continuous free phase layer upon the groundwater table. Due to density differences, dense non-aqueous phase liquids (DNAPLs) migrate further downward in the saturated zone, sometimes reaching great depth. These DNAPLs may also be present as pure product in the saturated zone.

Since the different mechanisms of NAPL movement and the dissolution are controlled by several factors, such as the chemical characteristics, the heterogeneity of the unsaturated zone and the aquifer, their presence is most of the time difficult to confirm. This results in a difficult quantification of the migration of the pollution. As a consequence NAPLs present in the subsurface may often be undetected, which results in incomplete site characterization followed by inadequate remedial actions.

To define proper remedial actions and for judicial reasons, correct characterization of the pollution (including NAPL) is necessary. A selection of detection and measurement techniques, which can be applied to identify and characterize a NAPL pollution, present in the soil, is discussed in the following sections. Some field results for an area, where a groundwater plume resulting from major spills of both LNAPL and DNAPL, are briefly discussed.

2. Selection of Detection and Field Measurement Techniques

For the characterization of NAPL pollution, field measurement techniques include both so-called conventional and alternative techniques. The application of conventional techniques is defined and described in regulations and protocols. such as the Flemish Decree on Soil Remediation. Conventional techniques include drilling and installment of observation wells, sampling and analysis of soil and groundwater (picture 1). These methods are necessary because they allow an actual view of pollution problems.

Since alternative techniques (e.g. geophysical, direct-push, soil gas sampling and field analytical techniques), in many cases, provide important additional information related to the soil quality, their application in environmental research has been encouraged by both the authorities and consultants.

Before planning any field work it is always necessary to collect all possible data related to the field situation. Only after interpretation of these data can a field investigation and sampling program can be set up. Special attention should always be given to the type of products used in the field and the identification of possible pollution source.

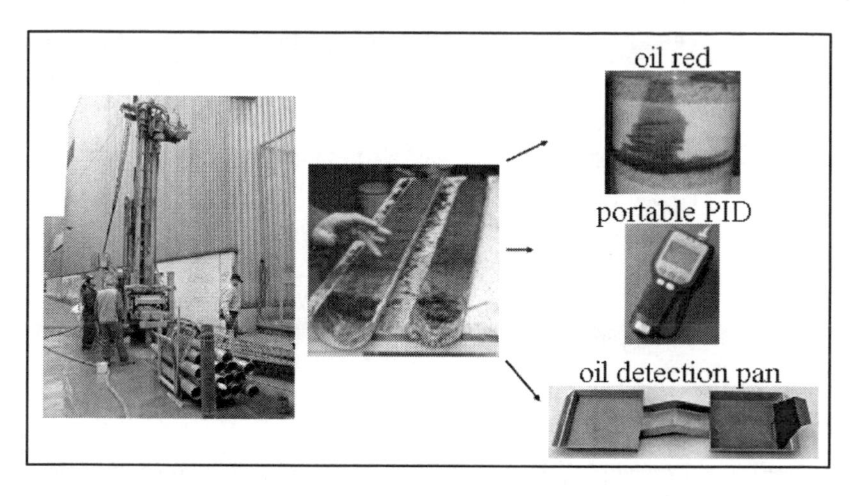

Picture 1 Collecting of undisturbed soil samples and overview of simple field methods such as the use of oil red, the oil detection pan and a portable PID to investigate the presence of NAPLs

2.1. SCREEN POINT GROUNDWATER SAMPLER

In addition to the observation wells the screen point groundwater sampler (SPGS) from Geoprobe Systems® (picture 2) can be used to take groundwater samples. The SPGS consists of a filter (length 1,2 m) connected to probe rods in

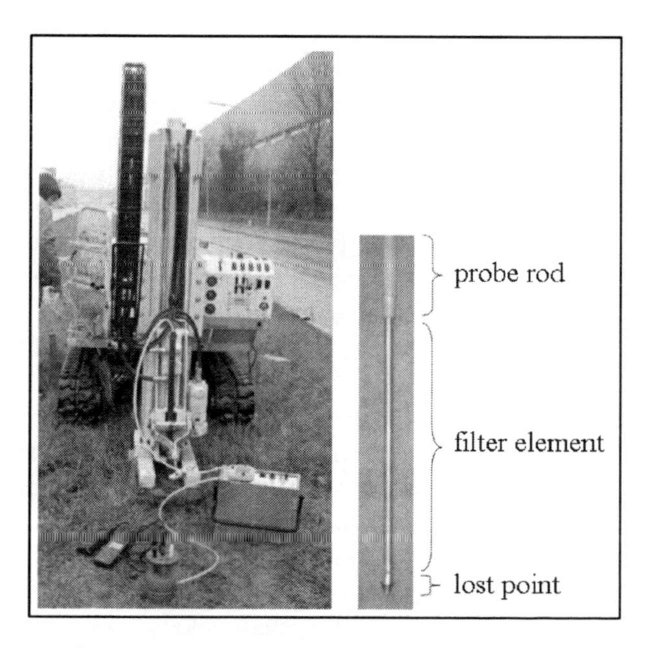

Picture 2 Illustration of the screen point groundwater sampler (SPGS)

which a tubing system is fixed to the filter. The SPGS is driven by direct push to the desired depth. Subsequently, the rods are pulled-up enabling the groundwater to migrate through the filter.

Advantages: sampling at discrete depths, on site decision, the possibility of vertical and horizontal screening and the lower installment costs (Bronders et al., 2000).

Disadvantages: the placement of the SPGS is temporarily, no re-sampling is possible, only limited data on groundwater levels can be obtained.

2.2. MEMBRANE INTERPHASE PROBE

The membrane interphase probe (MIP™), a screening tool developed by Geoprobe Systems®, can be used to obtain a semi-quantitative vertical concentration profile of volatile organic carbons (VOCs) present in the subsurface. Correlation of several MIP profiles allows for delineation of a VOC source zone, both in vertical and horizontal direction, with limited laboratory costs (Christy, 1996; Rogge et al., 2001).

In the field the MIP probe, which contains a semi-permeable membrane, is hammered or pushed into the ground (picture 3). The temperature of the probe ranges between 80° and 121°C and forces VOCs, present in the soil surrounding the probe, to vaporise. Subsequently the different volatile components extracted from the soil matrix diffuse through the membrane and are carried to a combination of three detectors (flame ionisation detector [FID], photoionization detector [PID] and dry electrolytic conductivity detector [DELCD]) by a constant flow of nitrogen gas. A continuous log of a wide range of aromatic and halogenated hydrocarbons versus depth may be generated. The detector response corresponds to the sum of the individual signals received for different individual pollutants. Since the MIP-probe is also equipped with a conductivity dipole, soil conductivity data are also collected at the same time.

The FID detects the presence of both aliphatic and aromatic hydrocarbons. Components with an ionisation potential < 10.2 eV (e.g. BTEX and some CAHs) are recorded by the PID. The DELCD is sensitive to the presence of chlorinated and brominated components.

Since the detection limit for BTEX components and CAHs present in the soil and/or subsurface are above most soil reference values applied in many countries the MIP-system can only be used for the delineation of source zones (including pure product).

Evaluation of the MIP application carried out by Bronders et al. (2000) showed that the application of the MIP system is very useful to obtain an in situ continuous vertical semi-quantitative VOC profile and to get an indication of

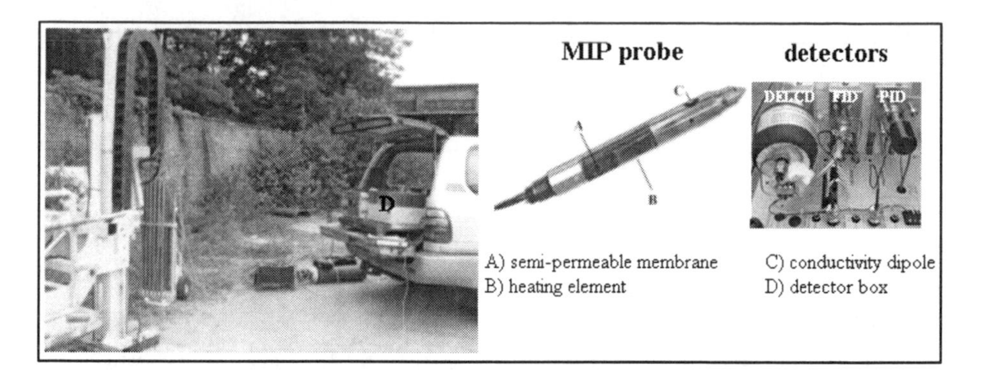

Picture 3 Application of the membrane interphase probe (MIP)

the presence of pure product. But when interpreting MIP results, reference analyses are always needed to obtain qualitative and quantitative results. Since detection limits are high MIP probing is only used to delineate source zones.

Advantages: continuous profile of information, detection of several VOCs (including mono - aromatic compounds and CAH), allows decisions on the field.

Disadvantages: semi-quantitative (when only using the detectors stand-alone); limited sensitivity (high concentrations of VOC have to be present), always need for conventional sampling and analyses, limited in depth.

2.3. COMPOUND-SPECIFIC STABLE ISOTOPE ANALYSIS

Compound-specific stable isotope analysis (CSIA) is being increasingly applied to investigate and monitor the sources, transport, and fate of organic contaminants in environmental systems including groundwater (e.g. Slater, 2003; Meckenstock et al., 2004). For example combined $^{13}C/^{12}C$ and $^{2}H/^{1}H$ analysis of monoaromatic hydrocarbons, in combination with conventional molecular fingerprinting, can be used to evaluate the impact of different sources on a contaminated aquifer and to evaluate the presence of degradation processes.

Results for carbon and hydrogen isotope compositions are reported in δ per mil (‰) relative to respectively the standards V-PDB and V-SMOW. Analytical precision is represented as the standard deviation of triplicate or duplicate measurements.

Advantages: allows distinction between sources, indication of biodegradation,

Disadvantages: relatively expensive, no standard method available, time consuming analyses.

3. Application

The industrial area of Vilvoorde – Machelen is located along the river Zenne at the northern edge of Brussels (Belgium). Soil and groundwater investigations carried out at 250 parcels have indicated the presence of a regional groundwater pollution (Figure 1).

Four distinct source zones were identified. Source area 1, resulting from the activities of a former paint and varnish factory, is characterized by the presence of a LNAPL pool. At source area 2 (a former solvent recycling site), and source area 3 (car manufactory) both LNAPL (BTEX) and DNAPL pure product were identified. Source area 4, located near a paint factory, contains high concentrations of DNAPLs. To characterize the pollution a combination of different techniques has been used.

Because of an ongoing dispute concerning the liability of remediation specific attention has been given to source areas 1 and 2. The results presented are obtained in the framework of this site specific research.

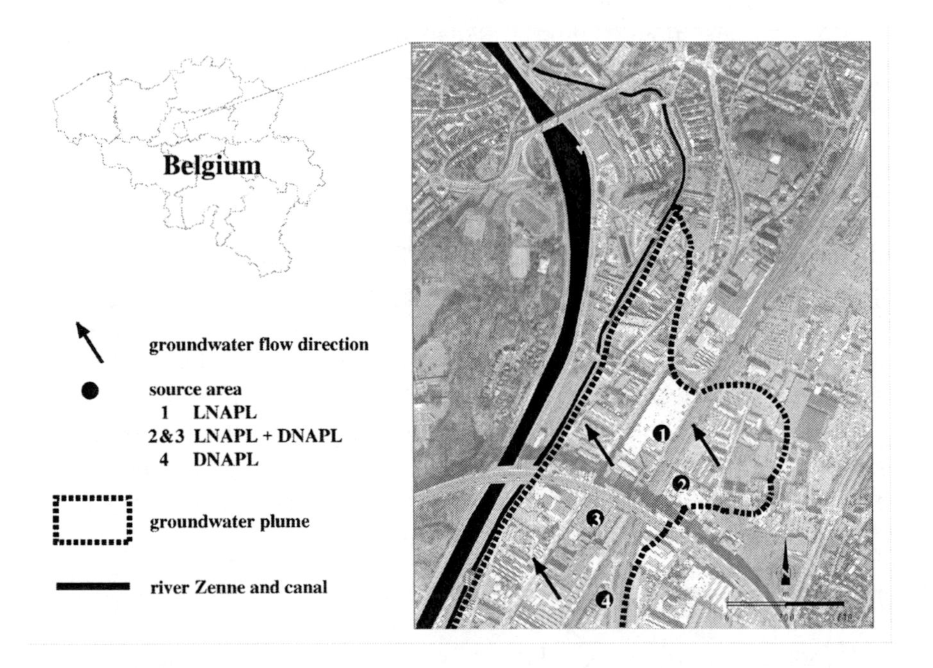

Figure 1. Overview of the location of the study area of Vilvoorde – Machelen

3.1. CONVENTIONAL TECHNIQUES

A region wide inventory of the soil and groundwater quality together with site specific research was carried out including drilling and the installment of observation wells. To prevent cross contamination the use of casings during drilling was mandatory. At a selection of locations, undisturbed samples were taken to describe the lithology in detail, to detect the polluted zones and to select samples for further analysis. Hereby oil red-o (a hydrophobic dye), oil detection pan and a portable PID (see also picture 1) were used to respectively illustrate the presence of pure product and give semi-quantitative information regarding the presence of volatile organic carbons. In conjunction, groundwater levels and the relative thickness of pure product was measured. Groundwater samples were taken for further analyses.

3.2. ANALYTICAL RESULTS

The LNAPL pure product present at source area 1 has a total BTEX concentration up to 120 g l^{-1}. In the shallow groundwater, < 8 m-below ground surface (bgs), concentrations of more than 10 mg l^{-1} were measured. Benzene and mp-xylene are the most abundant components in both the pure product and the groundwater.

Source area 2 is characterised by the presence of a pure-phase hydrocarbon layer containing both BTEX (total concentration about 140 g l^{-1}) and CAH (total concentration up to 17 g l^{-1}). In the shallow groundwater (< 8 m-bgs) contaminant concentrations up to 125 mg l^{-1} were detected. Also at greater depth (> 8 m), the groundwater quality is significantly affected. Along the groundwater flow path, all contaminants show a strong decline in their concentration with increasing distance form the source.

3.3. MIP PROFILES

The output from the MIP acquired near source area 2 and source area 1 are provided in figure 2 and figure 3, respectively. In both figures data with respect to the soil electrical conductivity (EC), the speed of penetration, the detector signals (FID, PID and DELCD) and temperature of the membrane are given versus depth.

At source area 2 the FID, PID and DELCD show similar profiles. High peak signals (intensity > 5 V) are obtained at ~4,5 m-bgs, between 9–10 m-bgs and from 13 m-bgs to ~14 m-bgs indicating high concentrations of CAHs. Because of the very high signal (50V) at 9–10 m-bgs, the presence of pure product is suspected.

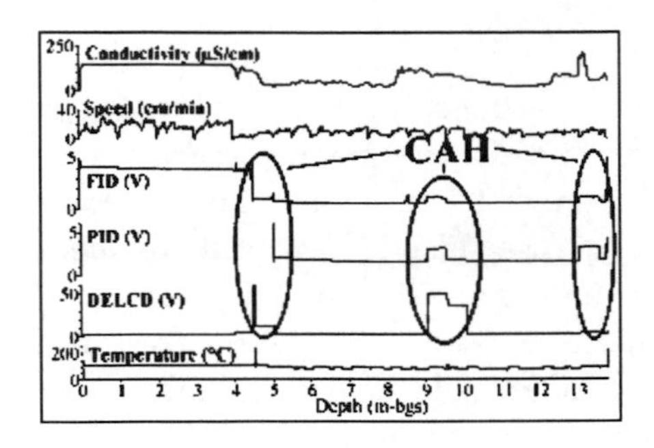

Figure 2. MIP output source area 2

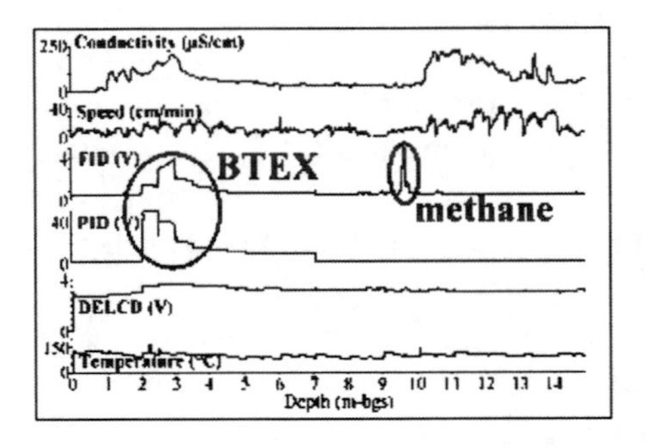

Figure 3. MIP output source area 1

Near source area 1, the signals recorded from 2 to ~7 m-bgs by the FID (maximum intensity 3,5 V) and PID (maximum intensity 40 V) detector reflects the presence of very high concentrations of BTEX components. The absence of any DELCD signal means that at this location no high concentrations of CAH are present. The isolated peak at the FID profile recorded ~10 m-bgs is explained by the presence of peat, which was confirmed by geological information collected in the area.

The EC profiles correspond to the lithological composition of the subsurface. At the top and the bottom of the MIP profiles high EC signals (100–200 µS/cm) reflect the presence of fine grained sediments (silty/clayey composition. Coarse grained sediments (sand and gravel embedded in a sandy matrix) result in lower EC signals (~50 µS/cm).

3.4. COMPONENT SPECIFIC ISOTOPIC COMPOSITION

Stable isotope results were obtained by a gas chromatography – isotope ratio mass spectrometrysystem (GC-IRMS, Delta^{plus} XP, ThermoFinnigan, Germany). For the carbon and hydrogen isotope analyses the GC-IRMS system was equipped with respectively a combustion interface and a thermal conversion interface. The stable isotopic analyses were executed at the Netherlands Institute of Applied Geoscience TNO.

The carbon and hydrogen stable isotopic composition of ethyl benzene/ m&p-xylene (eB-mpX) present in the pure product and in the "near source" groundwater samples of both source area 1 and 2 is given in figure 4. From this figure it can be concluded that:

- the $\delta^{13}C$ - $\delta^{2}H$ isotopic signature of eB-mpX from source area 2 are depleted compared to the isotopic composition of eB-mpX in source area 1;
- for both sources the isotopic signatures of eB-mpX in the groundwater "near source" samples are enriched compared to the isotopic composition of eB-mpX in the NAPL pure product.

These results indicate the possibility of isotopically distinguish the source zones and the presence of microbial degradation. The latter is supported by the presence of degradation products in the ground water samples.

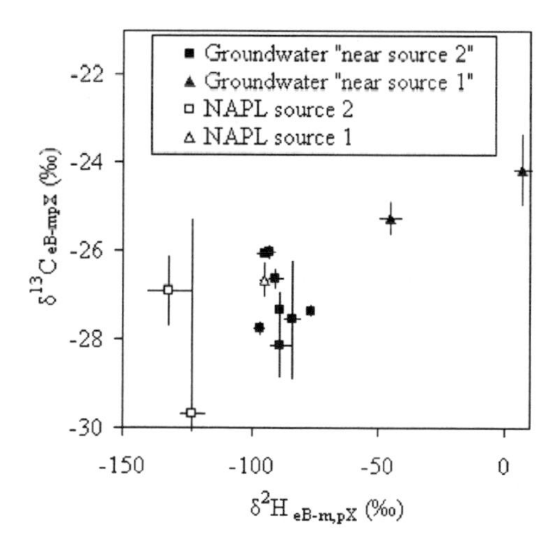

Figure 4. $\delta^{13}C$ versus $\delta^{2}H$ values for ethylbenzene – mp-xylene (eB-mpX) extracted from pure product (NAPL) and "near source" groundwater samples

To evaluate the influence of both sources on the overall groundwater contamination, the BTEX isotopic composition across the resulting groundwater plumes was determined. As illustrated in figure 5 the carbon isotopic composition of eB-mpX extracted from mainly the deep groundwater samples ($\delta^{13}C$ range from –26.8 to –25.5‰) are depleted compared to isotopic composition of eB-mpX extracted from mainly the shallow groundwater samples ($\delta^{13}C$ range from –24.7 to –22.8‰). Unfortunately these results were not confirmed by the isotopic composition of other BTEX components (Verhack, unpublished PhD results).

According to these results CSIA can be applied to make a distinction between different source zones and to identify the presence of biodegradation. Differentiation between mixing plumes however is very difficult.

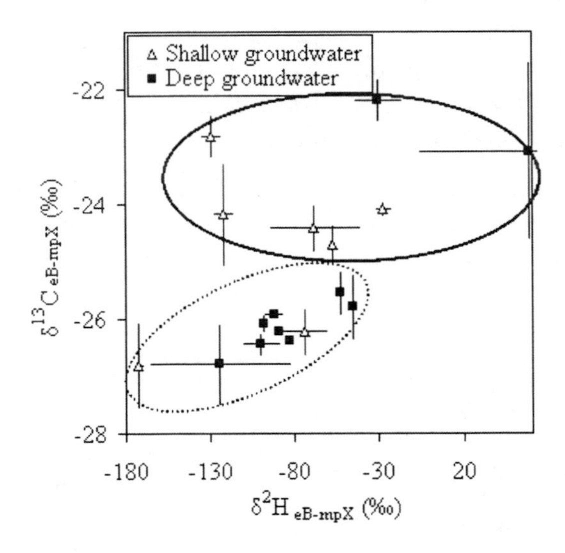

Figure 5. $\delta^{13}C$ versus $\delta^{2}H$ values for ethylbenzene – mp-xylene (eB-mpX) extracted from shallow (filter depth < 8 m-bgs) and deeper (filter depth > 8 m-bgs) groundwater samples

4. Conclusions

It is advised that at sites where NAPLs are present a combination of so-called conventional and alternative detection and field measurement techniques should be used. The combination of drilling, the placement of observation wells and screen point groundwater samplers, soil and groundwater analyses and the application of membrane interphase probing, component specific isotope analysis allows for a better characterization. Notice that when using "alternative techniques" the user should at least be aware of the limitations of the applied methods. In any case a proper desk study should be performed before starting any field and analytical campaign to select the most appropriate techniques.

Acknowledgement

This study has been supported financially by the Flemish Public Waste Agency (OVAM).

References

Bronders, J., Olivier, I., Vanermen G., Wilczek, D., Ceenaeme, J., and Wille, E. (2000), "Dense chlorinated solvents in groundwater: a methodology for characterisation", 7th Int. FZK/TNO Conf. on Contaminated soil, Germany, 326–327.

Christy, T.M. (1996), "A driveable permeable membrane sensor for the detection of volatile copmpounds in soil", 1996 National Outdoor Action Conference, Las Vegas Nevada.

Huling, S.G., and Weaver, J.W. (1991), "Dense Nonaqueous Phase Liquids", EPA Ground Water Issue, EPA/540/4-91-002, U.S. EPA, R.S. Kerr Environ. Res. Lab., Ada, Oklahoma, p. 21.

Meckenstock, R.U., Morasch, B. Griebler, C. and Richnow, H.H. (2004), "Stable isotope fractionation analysis as a tool to monitor biodegradation in contaminated aquifers", *Journal of Contaminant Hydrology,* 75, 215–255.

Newell, C.J., Acree, S.D., Ross, R.R. and Huling, S.G. (1995), "Light Nonaqueous Phase Liquids", EPA Ground Water Issue, EPA/540/S-95/500, U.S. EPA, R.S. Kerr Environ. Res. Lab., Ada, Oklahoma, p. 28.

Rogge, M., Christy T.M., and De Weirdt, F. (2001), "Site contamination: fast delineation and screening using the membrane interpphase probe, in Breh et al. (eds) Field Screening Europe 2001, Kluwer Academic Publisher, The Netherlands, pp. 91–98.

Slater G.F. (2003), "Stable isotope forensics – When isotopes work" *Environmental Forensics*, 4, 13–23.

IMMOBILIZATION OF HEAVY METALS AND STABLE ORGANICS FROM AQUEOUS SYSTEMS ON MODIFIED ACTIVATED CARBON

TUDOR LUPASCU*, MARIA TEODORESCU[2]
** Institute of Chemistry, Academy of Science of Moldova, 3 Academiei Street, MD 2028 Chisinau, Republic of Moldova.*
[2] National Research and Design Institute for Industrial Ecology-ECOIND, 90-92 Panduri Road, Sector5, 021621 Bucharest, Romania

Abstract. The negative impact of contamination with heavy metals and stable, hazardous organics, on the one hand and the presence of these contaminants in the human body, on the other hand have been the matter of an increasing concern in the last decades. Exploiting available, although limited resources for designing new methods designated to contaminants' immobilization without generation of additional risk was studied by a Moldavian – Romanian joint team. Active Carbons (AC) obtained from fruit processing waste, as nut shells and plum stones were oxidized to forms with significant ion exchange capacity. The Oxidized Active Carbons (OAC) were characterized and used to selectively adsorb heavy metal ions (as Cadmium – Cd, Copper – Cu and Lead – Pb) or hazardous organics (as ortho-Nitroaniline – ONA). The adsorbent capacity and selectivity was different when adsorbents of different origin were used and also when the adsorption process was applied to mono- or multi-component solution.

Keywords: oxidized active carbon, adsorption, ion exchange, adsorption capacity, selectivity

1. Introduction

The oxidized active carbons (OAC) are known to have a range of valuable properties. They are highly selective complexating ion exchangers, efficient adsorbents, active and selective catalysts of many reactions (L. A. Tarkovskaya and others, 1995).

M.D. Annable et al. (eds.), Methods and Techniques for Cleaning-up Contaminated Sites, 71–79.

The ability of OAC to adsorb cations of metals has been known since 1957 (M. M. Dubinin). It was demonstrated that the adsorptive activity in relation with metal ions is conditioned by the presence of a large quantity of functional groups of acid type on the OAC surface. These groups have different composition and abilities to get ionized. At the interaction of ions, carrying charge of two or three, with the surface of oxidized carbons, besides the exchange of ions, coordinative interaction followed by formation of surface complexes occurs.

The high selectivity of OACs, appearing in numerous systems, determines their successful application to different industrial processes for addressing economical and ecological problems as well as to medicine for adjusting acid-base equilibrium of organism, salt composition of internal medium, etc. (V. I. Davydov and others, 1993).

Besides their above-mentioned properties, OACs have other advantages, as thermal and chemical stability and regenerability (as described by I. A. Taikovskaya in 1981). Thus, the OACs are efficient materials for extracting heavy metals from contaminated aqueous systems.

The oxidized carbons are produced from usual carbons, including active carbons (AC) using treatment with different oxidizing agents, in gaseous or liquid form. The traditional oxidizing agents are nitric (V) acid, ozone, sodium hypochlorite, potassium permanganate, oxygen from air, etc. (process described by H. P. Boehm in 1994).

2. Experimental Work

The laboratory work was directed to:

a. obtaining OAC from fruit seed - based AC;
b. OAC characterization vs. the original AC;
c. adsorption study for heavy metal ions immobilization on OACs;
d. possible use of OAC vs. AC for organics immobilization.

2.1. OAC$_S$ PREPARATION

Active carbons, obtained through traditional methods from peach stones (AC-P) and nut shell (AC-0) were used as the initial material subjected to oxidation.

The oxidation was performed using 50% nitric (V) acid under constant heating (water bath) for 8 hours, using the ratio AC: $HN0_3$=50 g : 150 mL. The resulting samples of the oxidized carbons OAC-P and OAC-0 were washed with distilled water, then consequently treated with 0.1N NaOH (for removal of humic substances and remaining $HN0_3$), then they were washed-up until reaching weak acid reaction. Finally they were dried at 105°C to constant weight.

2.2. OAC$_S$ CHARACTERIZATION

The adsorption capacity, when applied to metal ions, is logically assimilated to ion exchange capacity. The *full exchange capacity* (EC) of OAC was assessed in the reaction with 0.1 N NaOH.

Qualitative and quantitative composition of the surface functional groups was determined by neutralizing them with bases of different strength: 0.05 N solutions of $NaHCO_3$, Na_2C0_3, NaOH and 0.1N HC1 (as Boehm described in 1994 and 1999).

Separation of the functional groups by basicity was carried out taking into consideration the postulate that the solution of $NaHC0_3$ neutralizes strongly acid groups, Na_2C0_3 neutralizes both strongly and weakly acid groups, NaOH neutralizes all carboxylic and phenolic groups.

In order to evaluate the adsorption volume of the OAC samples vs. the corresponding ACs, their adsorption capacities were tested using standard methods. Three chemicals were used: methylene blue, iodine and benzene (excication method). Ion exchange and adsorption characteristics were quantified.

2.3. STUDY OF HEAVY METALS ADSORPTION ON OAC$_S$

Adsorption of Cu, Cd and Pb ions from their 0.01-0.01 M nitrates (V) was studied under static conditions, at the ratio of adsorbent: solution 0.1 g: 25 mL.

Preliminary work was carried to investigate the adsorption kinetics. Although the results obtained demonstrated that the saturation of the oxidized carbons is reached in about 6 hours, the contact of the carbon with the solution lasted for 24 hours to reach equilibrium (standard contact time for adsorption isotherms). After this, the solution was filtered and the equilibrium concentration of ions of the metals was determined.

The equilibrium concentrations of $Cu^{2}+$, $Pb^{2}+$ and $Cd^{2}+$ were determined by the method of oscillopolarography with linearly changing voltage. Oscillopolarogrmmes were registered using an Oscillopolarographer P05122 model 03.

Value of adsorption of ions was calculated by the equation:

$$a = \frac{(C_o - C_e)V}{m} \tag{1}$$

where: a – value of adsorption capacity (meq/g); C_0 – initial concentration of metal ion (meq/mL); Ce – equilibrium concentration (meq/mL); V – volume of solution kept in contact and m – mass of Carbon used (g).

2.4. IMMOBILIZATION OF ORGANICS USING AC_S AND OAC_S

Ortho-Nitroaniline (ONA) was selected as stable, hazardous organic to be immobilized on AC and OAC. The exhausted OAC, after saturation with Pb^{2+} ions was also used as adsorbent.

3. Results and Discussion

The results obtained in the above mentioned experimental work encouraged considering the OAC usage on a larger scale.

3.1. MAIN CHARACTERISTICS OF AC_S VS. OAC_S

Oxidation of the active carbons leads to changes in some characteristics, in relation with both adsorption capacity and structure.

The oxidized carbons OAC-P (plums) and OAC-0 (nut shells) were characterized from the point of view of:

a. Ion exchange capacity (Table 1);
b. Distinguishing between acidic groups of different strength (Table 2) and
c. Adsorption standard characteristics (Table 3).

TABLE 1. Ion exchange characteristics of oxidized active carbon (OAC) and active carbon (AC)

Adsorbent	Ion exchange capacity (EC), meq/g, with				
	0.1 N NaOH	0.05 N NaOH	0.05 N Na_2C0_3	0.05 N NaHCOs	0.1N HCl
OAC-P	3.80	3.30	2.05	1.60	0.18
AC-P	0.80	0.70	0.30	0.20	0.54
OAC-0	4.20	3.40	2.00	1.65	0.27
AC-0	0.40	0.40	0	0	1.06

TABLE 2. Composition of active groups at the surface of OAC and AC

Adsorbent	Quantity of groups, meq/g		
	Carboxylic groups		Phenolic groups
	Weakly acid	Strongly acid	
OAC-P	1.6	0.45	1.25
AC-P	0.2	0.1	0.4
OAC-0	1.65	0.35	1.4
AC-0	0	0	0.4

TABLE 3. Standard adsorption characteristics of OAC and AC

Adsorbent	Adsorption capacity, meq/g		Specific volume, cm^3/g	Surface area, m^2/g
	by Methylene blue	by Iodine	by Methanol	by Methanol
AC-P	330	1231	0.457	900
OAC-P	165	810	0.445	875
AC-0	300	1239	0.4520	1009
OAC-0	210	950	0.456	972

Data given in Table 1 demonstrate that the original carbons (AC) contain a certain amount of functional groups. Anyway, it is to be noticed that the active carbon produced from the peach stones carries mainly acid groups, being thus a cation exchanger, while the active carbon produced from nut shell that carries mainly basic groups is an anion exchanger.

A great number of functional groups of different basicity form at the surface of active carbons during their oxidation with the final result in

- *spectacular increase of acidic group number and*
- *significant decrease of alkaline group number.*

The results presented in Table 2 show that the quantity of acid groups (strong, weak and phenolic) are reasonably comparable for both oxidized sorts of AC.

Data presented in Table 3 demonstrate that the AC oxidation results in decrease in adsorption characteristics: adsorption capacity (as determined using iodine) by 34% for OAC-P and by 24% for OAC-0. Adsorption capacity (as determined by methylene blue) is even more adversely affected: a decreased by 50% for OAC-P and by 30% for OAC-0 was established. These results prove that the porous structure of ACs was affected: part of micropores of active carbons become mesopores which, in their turn, are transformed into macropores. It should be also mentioned that the values of adsorption volume (Vs) and the geometrical surface (surface area, or specific area) of oxidized sorts, compared with the non-oxidized ones (as determined by the adsorption of methanol), although not significantly, are also adversely affected.

3.2. HEAVY MATAL IONS IMMOBILIZATION ON OAC_S

Although the adsorption kinetics (as shown in Figure 1) demonstrate that 6 h is enough time to reach adsorption equilibrium for all three metal ion, the contact was maintained up to 24 h.

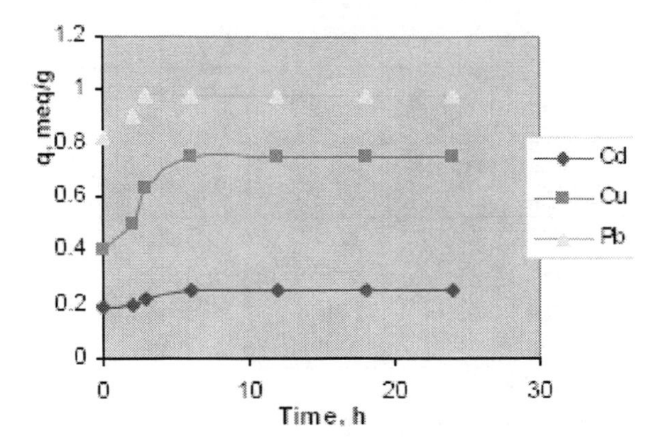

Figure 1. Kinetics of metal ions adsorption from 0.005 M Nitrate solution on Oxidizes Carbons.

Analysis of the adsorption isotherms for metal ions, when immobilized from mono-component solutions shows that the adsorption values for the three studied metals follow the series:

- Cd<Pb<Cu for OAC-P (Figure 2) and
- Cd<Cu<Pb for AOC-0 (Figure 3).

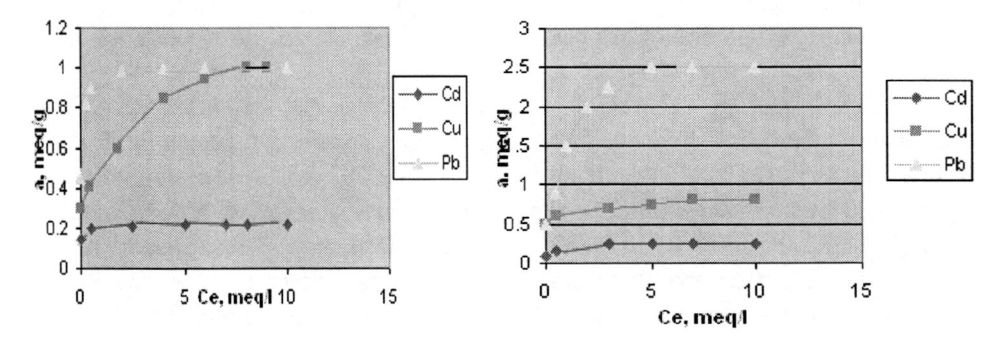

Figure 2. Adsorption Isotherms of metal ions on OAC-P, mono-component soln.

Figure 3. Adsorption Isotherms of metal ions on OAC-0, mono-component soln.

As noticed, values of adsorption capacity (meq/g) are more than double when using OAC-0 than OAC-P, but much lower than the static exchange capacity (EC) of oxidized active carbons (Table 1), i.e. max. 1.2 (for Cu and Pb), vs. 3.8, when adsorbed on OAC-P and max. 2.5 (for Pb), vs. 4.2, when adsorbed on OAC- 0, respectively. A possible explanation for this behavior is the non-availability of some acid groups due to steric barriers. Thus, utilization of 8% of the adsorption potential only was registered for Cd ions and 27% for Cd and Pb, when adsorbed on OAC-0 and 8% for Cd, 27% for Cu and 62% for Pb when OAC-0 was the sorbent.

Analysis of the adsorption isotherms for the same metals ions, when immobilized from their tri-component solution (as nitrates) shows a very interesting behavior:

- The OAC-P selectivity has changed in relation with Cu adsorption: this is worse adsorbed than Cd and the Cu adsorption decreases with the solution concentration increase. In the same time, although the Pb ions adsorbed on OAC-P (Figure 4) from the multi-component solution decreases by 30%, compared with the adsorption value for the mono-component solution, the total adsorption capacity of OAC-P is comparable (about 1 meq/g) with the maximum value for mono-component adsorption.
- A similar behaviour was also found at the adsorption on OAC-0 (Figure 5), except the more severe decrease (about 70%) in the adsorption of Pb ions, a smaller adsorption of Cu (25%), while the adsorption of Cd ions remains the same. It is to be noted that the overall capacity of OAC-0 is about 45% of the best value for the same adsorbent, when used fro mono-component adsorption. The same tendency was noticed in relation with Cu adsorption: the exchange capacity decreases with the solution concentration increase.

Such behaviour changes when the metal ions are adsorbed from their salt mixture are believed to be due to the ions interaction, both in solution and at the interface. This influences the ion ability to displace hydrogen ions in functional groups and also the stability of bounds of pre-adsorbed ions with different groups at the OACs surface.

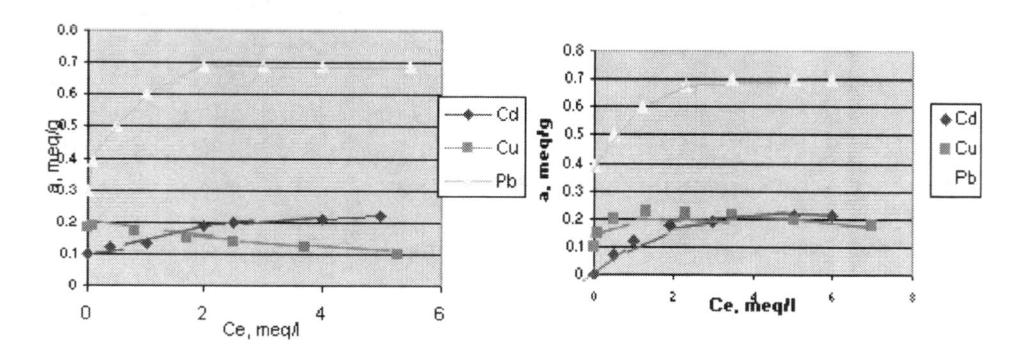

Figure 4. Adsorption Isotherms of metal ions on OAC-P, tri-component soln.

Figure 5. Adsorption Isotherms of metal ions on OAC-0, tri-component soln.

Studying the solution pH and conductibility changes showed that the metal ions immobilization is due in a higher proportion (about 70%) to a direct exchange with H^+ ions (from carboxylic groups) than to the coordinative interactions with oxygen atoms from the oxidized carbon matrix.

3.3. ORGANICS IMMOBILIZATION ON AC AND OAC

Ortho-Nitroaniline (ONA) was selected to investigate its immobilization on both original and oxidized AC (plum stones were the raw material for AC production). The oxidized sort was also used after its previous saturation with Pb^{2+}. The results (Figure 6) were quite contradictory to those obtained for metal ions immobilization.

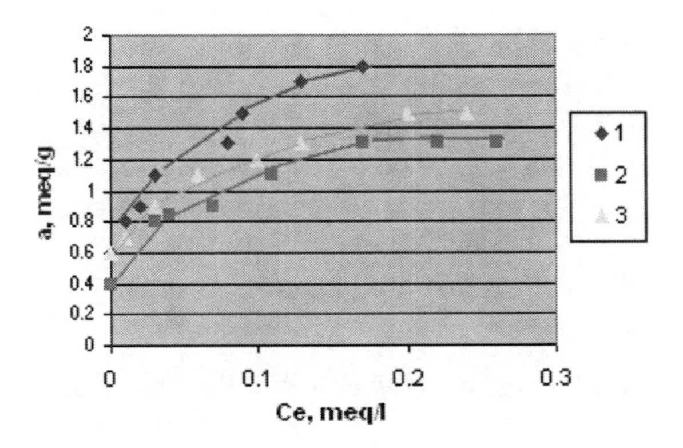

Figure 6. Adsorption Isotherms of ONA on AC (1), OAC (2) and OAC saturated with Pb^{2+} (3).

As shown, the AC adsorption capacity is slightly higher than that of OAC. This is due to the fact that ONA adsorption is driven by physical interactions than the ion exchange process. Changes of AC surface through oxidation, which generates an increased number of acidic functional groups, confer an increased hydrophilic character to OAC. This is not favourable to hydrophobic compounds adsorption. Also, the initial porous structure might have adversely been affected in relation with ONA molecules, too.

The relative increase of OCA adsorption capacity after saturation with lead ions is explained by the role of immobilized metal ions as newly generated active centres that attract (more than reject) the ONA molecule from the aqueous solution.

4. Conclusions

Immobilization of heavy metal ions from their aqueous solutions was successfully done using oxidized active carbons (OACs). The immobilization process was established to have been up to 70% a real ion exchange process than a physical adsorption one.

A significantly different exchange capacity (EC) for OACs was established when the ACs have different origin. The adsorption kinetics and equilibrium were studied vs. heavy metal ions (Pb, Cu, Cd) and hazardous organics (ONA).

An EC decrease of the oxidized sorts was established in relation with heavy metal ions and further decreased capacity was measured for sorption from multi-component solution, although the total capacity may be even higher than the EC itself.

The organic (Ortho-Nitroaniline - ONA) sorption was found to be governed by physical processes and less by chemical/exchange ones and the pre-existing metal ions on the sorption support were in favour of the adsorption process.

Corroborating different results, sometimes apparently contradictory, sustain the fact that every real context needs separate investigation and there are less common rules than factual differences.

Acknowledgement

This paper draws on the joint work done by Physico-chemical Laboratory of the Academy of Science of Moldavia, which carefully prepared the fruit-stone-based active carbons and characterized the adsorbents and to whom the authors are indebted. The work was partially supported by Governmental funds awarded for Academic or National research programmes.

References

L. A. Tarkovskaya and S. S. Stavitskaya. //Ros, *Him. Jurnal Him*., Obsch. D.I. Mendeleeva, 39, 44 (1995).
M. M. Dubinin, "*Poverhnostnde himiceskie soedinenia i ih roll v iavleniah adsorbtii*", Ixd.-vo MGU, Moscow, 1957 p. 9–33.
V. I. Davydov, S. S. Stavitskaya, V. V. Streiko and N. I. Kartel, "*Anterosorbtia: sostoianie, problemd i perspecnvd primenenia*", Naukova Dumka, Kiev, 1993, p. 67.
I. A. Taikovskaya, "*Okislenndi ugoli*", Naukova Dumka, Kiev, 1981, p. 200.
H. P. Boehm, *Carbon*, 32, 759 (1994).
H. P. Boehm, "*Stereohimia i mehanizmd organiceschih reac*[illegible]", Mir, Moscow, 1968, p. 168–235
Received: March 30, 1999 .

REMEDIATION OF THE FORMER MILITARY AIRPORT: TRIANGLE ZATEC

ROBERT RASCHMAN, JAN VANEK
Dekonta, a.s., Volutova 2523, 158 00 Prague 5, Czech Republic
Dekonta, a.s., Volutova 2523, 158 00 Prague 5, Czech Republic

Abstract. The paper refers to remediation of the former military airport premises, which is going to be reconstructed into a strategic industrial zone called TRIANGLE. The military airport was put into operation before the World War Two. During operation, petroleum hydrocarbon substances were stored and used on its premises. The oil products contaminated 120,000 m^3 of soil above groundwater level (unsaturated zone) and 100,000 m^3 of soil and underground water in zone of groundwater level and below it (saturated zone).

The objective of the remedial work is to remove contamination and to prevent further contamination spreading. Remedial activities started by additional remedial investigation (from August to September 2003), and was sequentially followed by *ex-situ* remediation of the unsaturated zone (from September 2003 to June 2004). During the *ex-situ* phase, 118,382 m^3 of soil contaminated above the target limit were excavated, biologically decontaminated and backfilled. In total, 411.4 tons of petroleum substances were removed. Furthermore, 64,612 m^3 of inert materials with contaminant content below the limit were excavated.

Presently, *in-situ* bioremediation of the saturated zone is being carried out. This part of remediation work should proceed till the end of the year 2008; post-remedial monitoring being planned till 2013.

Keywords: petroleum hydrocarbon substances, groundwater, soil, remediation, biodegradation

1. Introduction

The premises of the former military airport of Zatec are situated along the main road no. I/7 of Chomutov-Prague, near to the crossing with road no. I/27, Most-Pilsen, in the north - west part of the Czech Republic. In total, it has an area of

M.D. Annable et al. (eds.), Methods and Techniques for Cleaning-up Contaminated Sites, 81–90.

363 hectares and it belongs to typical "brownfields". Presently, these premises have been prepared for entry of a strategic industry called TRIANGLE (a strategic investor being in question) and the work was ordered by the area owner -Usti Region - in cooperation with the Investment and Development Agency (Czech Invest).

2. Site History and Objectives of Remedial Work

Zatec military airport was already introduced into operation before the World War II. Some equipment and technologies (mainly underground tanks of jet fuels), dating from that period, were used during the whole period of airport operation. Further enlargement of the airport took place during the War and also in the 1950's and the 1980's. Therefore, most of the equipment and technologies for storage and transport of the air fuels originated in the 1950's and were in operation up to the airport closure. The airport was shut down in 1992, but the army left the airport premises for good in 1997.

During the airport operation, the petroleum hydrocarbon based substances storage and usage on its premises caused the underground environment contamination, groundwater including,, particularly due to careless handling throughout the fuel delivery, distribution and usage. So, the present contaminants are mainly fuels (jet fuels – kerosene, oil, and petrol), light fuel oil, engine, transmission and hydraulic oils.

Extensive environmental contamination of the Zatec military airport premises was confirmed for the first time in 1980's, when investigations of similar sites and equipment were performed in former Czechoslovakia. In the case of Zatec airport, the investigation results confirmed fuel releases not only from underground tanks but also from the repumping equipment and distribution pipelines. The oil products contaminated approx. 120,000 m^3 of soil within the area of approx. 10 hectares above groundwater level (unsaturated zone) and the ground water within an area of approx. 30 hectares in the zone of groundwater level and below it (saturated zone).

The first remedial work started before the end of the 1980's, when the oil substances were pumped from groundwater level through hydrogeological wells located inside and outside of the premises. This was, in principal, an emergency solution to protect water resources situated near the neighboring villages called Nehasice and Tatinna against oil substances infiltration.

Systematic remediation of the premises started in August 2003, active operation of remedial technologies is designed till 2008 and post-remedial monitoring is planned till 2013. Remedial work is carried out in accordance with the decision of an appropriate administrative body (i.e. the Czech Environmental Inspectorate, Regional Inspectorate of Usti Region), which stipulates conditions for the remedial action, including the target limits for soil contamination (2,000 mg TPH per kg of dry matter) and for underground water (4 mg Total Petrolium Hydrocarbons - TPH per litre in contaminated plums, after removing the oil substances as non-aqueous phase). The Regional Inspectorate set the target limits based on the conclusions of risk analysis, ecological audit and some other studies (additional investigation, reports).

The estate owner and, at the same time, the beneficiary of the preliminary work for TRIANGLE industrial zone development is Usti Region.The general contractor for the remedial action is DEKONTA a.s. company. The supervising organisation, which controls the quality of the realized works, is AQUATEST a.s. company.

The objective of the remedial work is to remove contamination of underground environment and underground water in the premises of TRIANGLE industrial zone and to prevent further spreading of contamination via underground water flow.

3. Extent of Contamination

The remedial work started with additional remedial investigation in order to obtain the necessary data on the contamination extent of unsaturated zone, the underground body in both vertical and horizontal directions. These data were used mainly for specification of soil excavation areas but also for the optimisation of remedial technology parameters designed in the detailed remedial project.

In total, 57 proving boreholes were set out and drilled. The exact location of particular boreholes was decided by the results of all the investigation work and laboratory analyses concerning the TPH content in the soil samples. Detected contamination (kerosene mainly) did not exceed the level of 7,000 mg TPH per kg dry matter, but it should be pointed out that the wells were situated rather in border areas.. Contamination at level highly exceeding the target limit (up to 12,000 $mg.kg^{-1}$) was confirmed essentially in centres of considered excavation are, besides the site identified as the "Kitchen 1, 2". Other places with high concentration of contaminants were found e.g. in the northern area of the "Bunkers BLPH 3, 4" and in the western and northern parts of the "Laboratory LPH". The area of contamination extension did not differ too much from the data stated in the tender documentation. Contamination was also detected in soils of the saturated zone. These contaminated locations were identified by additional investigation e.g. in the western part of considered excavation on the "Laboratory LPH"site, in the southern part of considered excavation on the "Bunkers BLPH 1, 2" site and in the centre of considered excavation on "Bunkers BLPH 3, 4" and the 'Repump Station'sites.

The underground water level was detected at the depth of approx. 6 m below the existing ground level.

Based on the information obtained on geological profile, contamination and underground water levels, specific techniques of contaminated soil excavation with sequential decontamination were proposed for the following partial sites:

- "Bunkers BLPH 1,2" (excavation area of 10,792 m^2)
- "Bunkers BLPH 3,4" (15,743 m^2)
- "Repump Station" (5, 872 m^2)
- "Laboratory LPH" (2,105 m^2)
- "Main Pumping Station" (6,902 m^2)
- "Petrol Station PHM" (1,927 m^2)

4. Results of the Remedial Work and Discussion

Remediation of soils above the underground water level was carried out by excavation (Fig. 1) of contaminated soils and their sequential *ex-situ* biodegradation. Biodegradation was performed on a protected area (the former runway surface). After completing the remedial process (achieving the target limits for soil contamination) the cleaned soil was used as excavation backfilling.

After the excavation, the remediation of saturated zone, i.e. of the underground body below the underground water level, was continuously in-situ carried out. *In-situ* remediation was chosen, in this case, for the following two reasons: technical difficulty and high financial costs of contaminated soil excavation below underground water level and –also due to the fact that the contaminated area is partly placed below the road no. I/7 (Chomutov – Prague), where excavation could not have been performed. The technology of *in-situ* remediation is in principal the same as the *ex-situ* method except the fact that bacterial suspension and other additives are applied directly through infiltration drains into an underground body; the removal of soil is not carried out.

The hydraulic remediation - treatment of underground water - is closely connected with the technology of *in-situ* remediation. Underground water is pumped by a network of 30 wells located northern from the former airport premises, in a parallel line with the road no. I/7. This well system ensures the functioning of a protective barrier (thus preventing the contamination spreading along the ground water flow direction), as well as of the remediation process (the water pumped from the wells is treated up to the target limit and then infiltrated into the saturated zone through infiltration drains, at the bottom of remedial excavations).

Figure 1. Excavation work

4.1. EX-SITU SOIL REMEDIATION

Excavated soils from the unsaturated zone, which were contaminated above the target limit, were transported to the biodegradation field by lorries. The soil excavation was carried out from September 2003 till December 2003 and the runway situated on the premises of the industrial zone was used as a suitable site for biotechnological treatment of contaminated soil.

Ex-situ biodegradation was carried out in accordance with an approved operating procedure of the remedial design project. Before putting the runway into operation, the repair and control of technical security (clearness of collecting channels, leak-proof of discharge sewerage) of this site were performed. Simultanously, the reconstruction of a skimming tank (for rainfall water collection) was done. The size of the decontamination field was 80 × 2,500 m and its capacity was approx. 200,000 t of soil. This capacity was sufficient for completion of the remedial task in all to-be-cleaned areas of the airport premises.

After *ex-situ* decontamination, the runway demolition started with the last part of unnecessary building and object removal. This was carried out on the premises of the former military airport, from 2002 to 2004. Presently, the major part of the runway is demolished and the remaining part serves as a transportation infrastructure for further construction equipment.

During the biodegradation process, the runway area was divided into 10 deposit sectors and each of this further split into two parts - northern and southern. The excavated contaminated soils from individual decontaminated areas were temporarily deposited in these sections. The sections were marked and separated from each other in order to ensure the necessary handling area and technique (cultivators, cisterns, lorries). Soil of approximately the same contamination characteristics and content was placed in every sector. The objective of this set-up was to achieve the most effective process of *ex-situ* remediation in the sense of process optimisation (e.g. the retention time). The excavated soil was distributed on the site so its full aeration was ensured (average soil height was 90 cm). After verification of the achieved TPH limit (control sampling was done by the supervising organization, Fig. 2) the decontaminated material was sequentially transported back into excavated areas for backfilling.

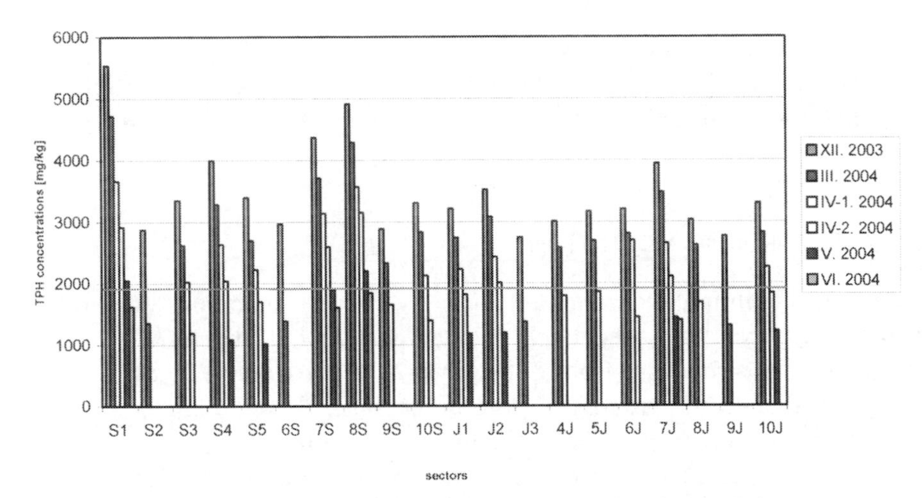

Figure 2. Soil TPH concentration during ex-situ biodegradation

The principle of biodegradation method is based on the ability of certain bacterial strains to utilize the existing contaminant (in the case of Zatec airport it is jet fuel – kerosene) as carbon and energy source in oxidation processes. Environmentally undesirable hydrocarbons (in case of kerosene, they are mainly aliphatic alkanes – linear and also branched, cycloalkanes, aromatic hydrocarbons and alkylated aromatic hydrocarbons) are decomposed following a standard metabolic pathways, through several intermediate stages, to carbon dioxide and water. Soil cultivation (loosening) was performed twice in the total volume of the treated soil during biodegradation process. Soil cultivation was carried out due to three following reasons: material aeration, uniform distribution of biopreparation and/or additives and material homogenisation. Cultivation was firstly performed by re-ploughing and soil shifting by scoop-loaders, in the second stage.

After finishing the excavation work, the bottom of remedial excavations was made into downhill grade, in direction of the underground water flow, and loosened to ensure an infiltration area for the applied biopreparation solution and mineral enrichment. Drainage distribution systems were installed on prepared bottom. At the same time, a central delivery system of the biopreparation solution, drawn from a technological centre to individual remedial excavation areas, was laid down. Infiltration drains were backfilled (recycled concrete, 10 mm grain size) up to 0.5 m thickness.

After performing the preliminary *in-situ* remedial work, the excavated zones were backfilled with decontaminated soil from the biodegradation field. Backfilling was performed building 0.5 m – thick – layers, which were compacted by a vibrating roller. This work was finished in December 2004.

During *ex-situ* remedial work 118,382 m^3 of contaminated soil were excavated, decontaminated and backfilled. Necessary inert materials (with contaminant content below the limit) were excavated in volume of 64,612 m^3. Within the framework of preliminary work (carried out before the beginning of the own *ex-situ* remediation) 374 m of steam piping and 2,909 m of sewerage were eliminated,, together with reprocessing of 14,673 t of wrecking rubble and 3,344 t of concrete resulting from the demolition work. At the same time other systems i.e. water piping, heavy- and light-electric power cables were also removed.

Within the soil decontamination work of the former runway 3,474 m^3 bacterial preparation, 421 m^3 liquid fertilizers and other additives were applied. During the remedial work monitoring and management systematic samples were collected and analysed: 4,877 soil samples and 283 underground water for determination of petroleum substances, 1,137 soil samples for biodegradation parameters and 471 samples of bacterial preparations for microbial quality.

Summing up, a total of 411.4 t of petroleum substances were removed during the *ex-situ* soil remediation.

4.2. IN-SITU SOIL REMEDIATION

The preliminary work of *in-situ* remediation consisted on laying down a 1,842 m delivery system for biopreparation distribution and 1,956 m of infiltration drains. 29 hydrogeological monitoring wells and 30 large-dimension remedial wells were further built. 119 ground water samples from the former as well as the newly constructed monitoring wells were drawned and analysed.

The monitored parameters, for this paret of work, were the followings: total petroleum hydrocarbons (TPH), content of dissolved oxygen, redox potential, temperature, conductivity and pH of water, content of nitrates and phosphates, data on chemical and biological oxidation demands in water samples and final bacterial activity. The obtained results were used within the optimisation of the applied biodegradation method.

Figure 3. Biodegradation technological centre

Before starting the *in-situ* biodegradation work the application of fertilizer solution (called "mineral enrichment") was necessary. As result, the maximal intensification of the naturally occurring microflora activity was achieved. The intensification results of the controlled dosage of mineral nutrients and micro-elements necessary for degradation activity performed by autochthonous; optionally introduced microorganisms. The composition of the applied fertilizer solution allows continuation of microbial metabolisms in anoxic conditions, in the case of limitation of air oxygen, in low permeable parts of the underground body. The solution is prepared in the biocentre (Fig. 3), in retention tanks, by diluting the liquid fertilizer with pre-treated ground water (taken from the well system of the hydraulic barrier). Mixing ensures complete dispersion of contingent sediment. Afterwards, the solution is delivered via pumping through the pipeline system into the infiltration drains, to the appropriate remedial excavations.

The *in-situ* method, as such, solves problems of re-treatment of the saturated zone contaminated with biologically treatable contaminants. It directly comes from the *pump&treat* technology and integrates biological elements into this classical technology. The method principle is based on optimisation of biodegradation process conditions, mainly in saturated zone of an underground environment. It uses natural biodegradation activity of autochthonous micro-flora as well as directly isolated bacterial strains, which are delivered into the underground environment. Aeration of pumped water and its enrichment with deficient nutrients (mainly nitrogen and phosphorus) and/or other way of electron acceptor dosage are applied for their activity stimulation.

Regular application of bacterial preparation started in November 2004 and it has been carried out, via infiltration drains, since then.

The *in-situ* remediation work will still continue up to the end of 2008 and, periodically, the quality of ground water will be observed up to 2013, within a post-remedial monitoring process in the whole premises.

4.3. PROTECTIVE REMEDIAL PUMPING AND UNDERGROUND WATER

Groundwater has been pumped from all 30 hydrogeological wells, which are forming the outside hydraulic barrier. The extracted groundwater is being treated in three containerized wastewater treatment stations. Both free phase of petroleum substances and dissolved organic contaminants are removed in these stations.

The remedial facility (station) is standardised and each consists of the following parts: a retention tank of pumped treated water with a support, a gravitation-coalescence separator, a labyrinth adsorption tank, a discharging pump, automatic control and electrical equipment, connecting pipeline, pipe fittings and water meters.

The treated ground water is used for production and dilution of bacterial preparation utilised within the remediation process. The *in-situ* technology is then a closed cycle of derground water: pump – treat – back infiltration. Due to this method, flushing of contaminated soils in the saturated zone is accelerated. The *in-situ* remediation has minimal demands terms of space requirement and thus it enables construction of communications, engineering systems and other equipment on the premises.

Within the remedial pumping process, approx. 46,000 m^3 of contaminated ground water is annually pumped and further treated from the hydraulic barrier wells.

5. Conclusions

The presented remediation case studyof the Zatec military airport is one of the largest remedial actions in Central and Eastern Europe.

During airport operation petroleum hydrocarbon substances stored and used on its premises caused massive pollution of 120,000 m^3 of unsaturated zone and 100,000 m^3 of saturated zone.

Systematic remedial activities of the site started with additional remedial investigation in August 2003. It was further followed by *ex-situ* biological remediation of the unsaturated zone, which was successfully finished in June 2004. During this phase 118,382 m^3 of soil contaminated above the target limit were excavated,, biologically decontaminated and backfilled. 411.4 tons of petroleum substances were removed.within the process.

At the moment, the saturated zone *in-situ* bioremediation is being carried out. This part of remediation work should proceed up to the end of 2008 and post-remedial monitoring is planned up to 2013.

AQUIFER REMEDIATION AND CHEMICAL RECOVERY FOLLOWING A SPILL DUE TO AN EARTHQUAKE IN TURKEY

CANER ZANBAK
Turkish Chemical Manufacturers Association,
Degirmen Sokak No: 19/9, Kozyatagi, 81090, Istanbul, Turkey

Abstract. In the aftermath of the August 1999 earthquake in Turkey, ruptures in piping connections of storage tanks resulted in an acrylonitrile spill. A post-incident risk assessment revealed that the acrylonitrile concentrations in the seawater and sediments rapidly decreased down to non-detect levels within a month of the incident. Periodic monitoring data revealed that the acrylonitrile contamination was limited to on-site soil and shallow groundwater at the site. A shallow groundwater monitoring and extraction program was initiated in October 1999. Groundwater had been pumped out of 3 manholes and 4 drainage ditches installed on-site. Acrylonitrile concentrations in one of the monitoring wells decreased from an initial high of 80,000 ppm down to non-detect levels at the end of the third year. This paper presents the hydrogeological setting, chemical recovery efficiency and successful outcome of a long-term environmental project on groundwater quality monitoring and shallow aquifer remediation following a chemical spill.

Keywords: chemical spill, aquifer remediation, groundwater extraction, chemical recovery

1. Introduction

The acrylic fiber manufacturing facility is located on the southern coast of the Marmara Sea, approx. 20 km. west of city of Yalova, Turkey (Figure 1). The facility has been in operation since early 1970s. Bulk chemicals are stored in a tank farm located on the coastline.

In the aftermath of the 17 August 1999 earthquake, three of the six on-site storage tanks were damaged at an acrylic fiber manufacturing facility and approximately 6,500 tons of acrylonitrile (ACN) was released into the environmental media due to ruptures in piping connections. No fire and fatal injury

M.D. Annable et al. (eds.), Methods and Techniques for Cleaning-up Contaminated Sites, 91–101.

occurred at the incident, which was directly attributed to the company's effective implementation of the Responsible Care® initiative of the world chemical industry, which was acknowledged in an ICCA-UNEP report submitted to the World Summit on Sustainable Development in Rio de Janeiro (2002a).

Majority of the spilled ACN liquid was released into the sea, whereas the rest seeped into the soil and vaporized. ACN, a chemical of high volatility potential at ambient conditions, readily biodegrades in the environmental media when diluted. A risk assessment conducted by the Turkish government agencies and research institutions revealed that the ACN concentrations in the sea water and shallow sediments rapidly decreased down to non-detect levels within a month of the incident, and the surface water sources, the vegetables and fruits in the nearby fields were not impacted. Cyanogenic impact of the chemical vapors was limited to a 200-meter radius of the tank area, as observed on browning of leaves of the shrubbery on-site (2001). There has not been reported a negative impact of the incident on human health to date.

Figure 1. Facility Location

2. Remedial Investigation Program

The facility is located on relatively compacted lower segment of a Pliosen age alluvial fan. Based on the available geotechnical borehole information, the tank farm area consists of a 4-6 meter-thick compacted fill and silty/sandy layer underlain by a min. 5 meter thick plastic clay layer overlain an another silty/sandy material layer throughout the length of 20 meter borehole logs.

The facility management initiated a long-term environmental investigation program immediately after the incident in coordination with the Turkish Ministry of Environment. The objective of the investigation was to determine the nature and extent of potential contamination in the sea, surface water and ground water and the surface soil and sediment quality.

The subsoil investigations in the vicinity of the tank farm revealed that the extent of soil contamination was vertically contained within the surficial 1 to 2 meter-thick compacted fill material.

Presence of the consolidated clay layer provided an aquitard barring the contaminated groundwater migration to the deeper aquifer as demonstrated by the absence of any ACN in the water samples collected from the facility's deep production wells. Lateral extent of this shallow groundwater contamination was also limited to facility boundaries due to presence of two water channels on the site boundaries providing natural hydrogeological barriers. A conceptual hydrogeological setting of the tank farm area is presented on a block diagram in Figure 2.

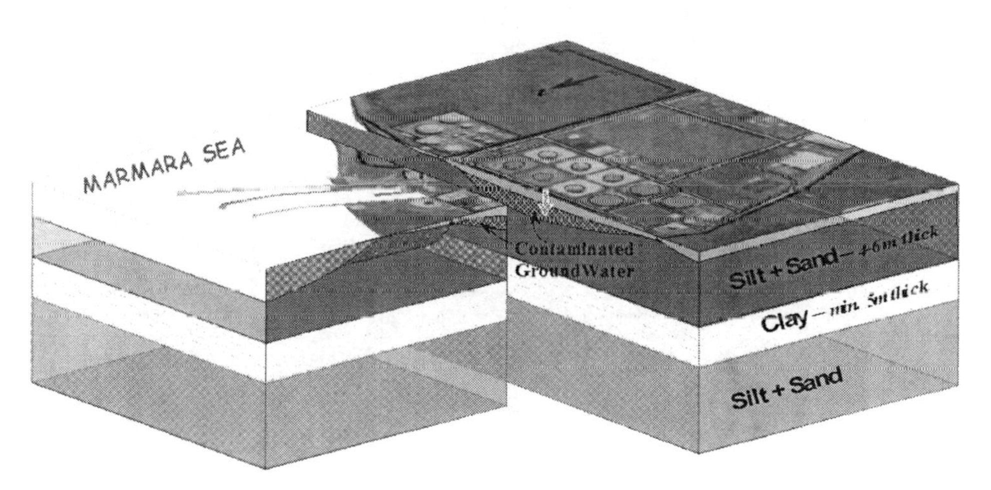

Figure 2. Conceptual hydrogelogical setting of the site

A shallow groundwater quality monitoring and extraction program was initiated in October 1999. Groundwater was monitored at 24 locations and pumped out of 4 installed drainage ditches *(french drains)* and 3 manholes installed on-site. The extracted groundwater was treated for acrylonitrile recovery at the facility's solvent distillation unit and then sent to the facility's wastewater treatment plant.

2.1. GROUNDWATER MONITORING PROGRAM

Shallow groundwater quality had been monitored for ACN, pH, temperature and specific conductivity parameters at 24 monitoring locations for seven years, between September 1999 and September 2006. Sampling frequency was bi-weekly for two months, weekly for 3 years, bimonthly for 2 years and monthly for the last two years. Groundwater monitoring locations are presented in Figure 3.

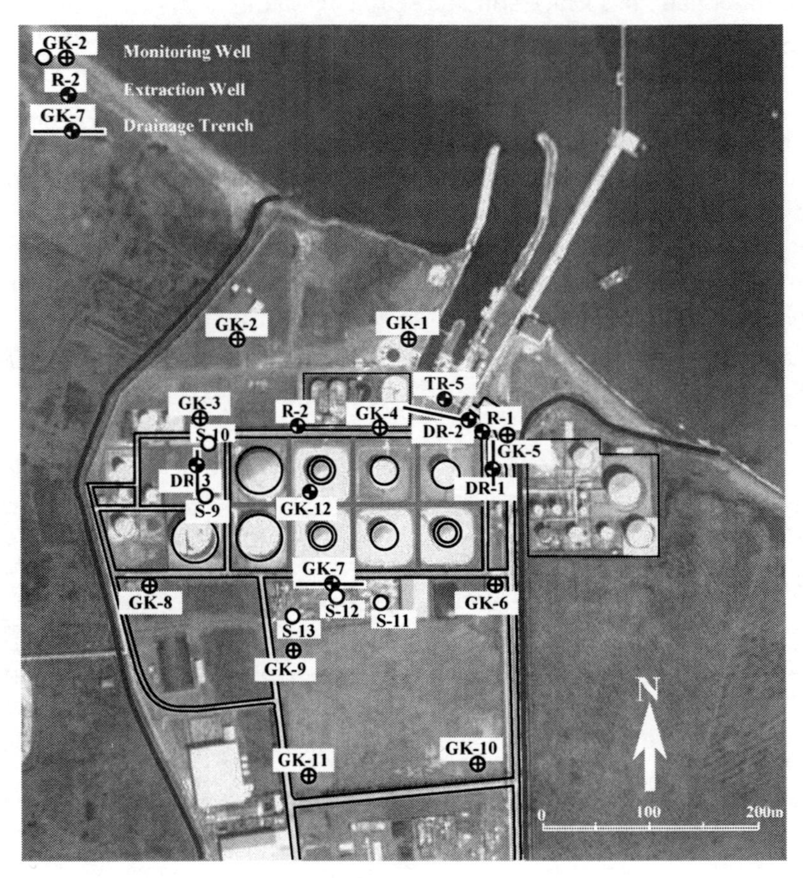

Figure 3. Groundwater quality monitoring and drainage trench and extraction well locations

Soil and groundwater sampling results at several offsite locations in the vicinity of the tank farm area revealed that the ACN contamination was limited only to on-site soil and shallow ground water in the vicinity of the tank farm area. No residual surface and sea sediment contamination was detected.

At the onset of the project, 11 monitoring wells (15 cm diameter steel casing) were installed (GK-series). Additionally, 5 existing geotechnical borings were converted to uncased monitoring wells by plugging the bottoms in the clay layer (S-series).

In the first month, three existing sumps located downgradient of the tank farm (R1, R2 and TR5) were used as the extraction points for groundwater. At the end of the first month of the project, one standalone extraction well (GK-12) and three drainage trenches (DR-1, 2 and 3) and GK-7 monitoring well were also converted to a drainage trench to expedite extraction efficiency. Typical construction details of the extraction wells are presented in Figure 4 (a) and (b).

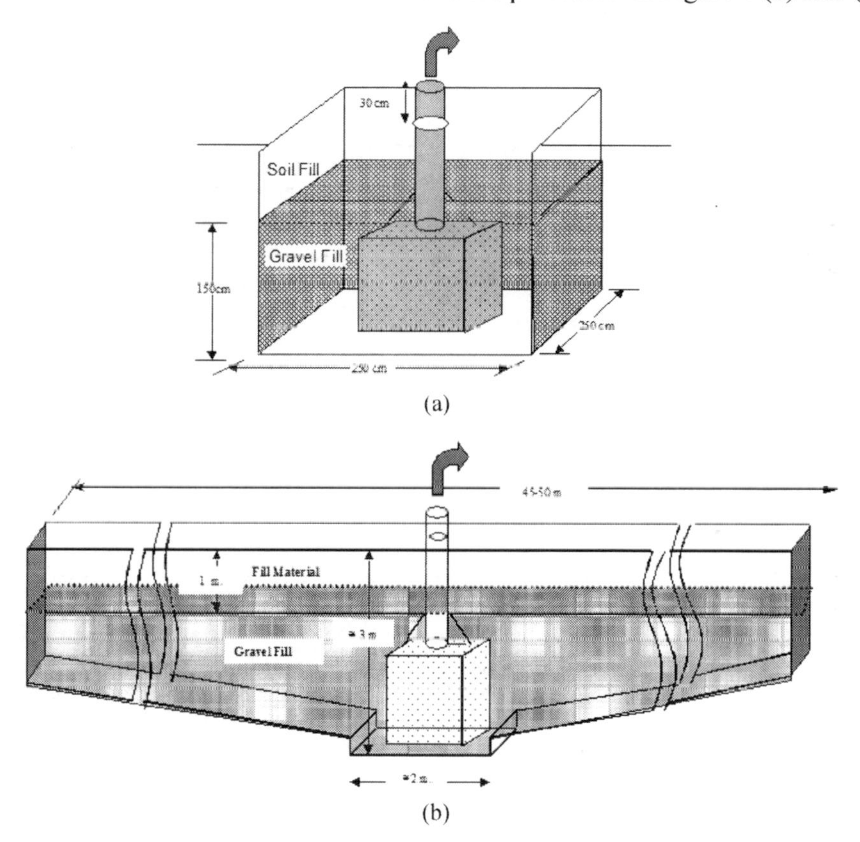

Figure 4. Typical details of (a) extraction well point and (b) drainage trenches (french drain)

2.2. GROUNDWATER EXTRACTION AND TREATMENT

Submerged pumps with level-switches were installed at each extraction point. Extraction of contaminated groundwater started at three facility drainage sumps located downgradient of the spill area at a total daily average of 35 m^3. This quantity was increased to an average 80 m^3/day after installation of the other drainage trenches. Extracted groundwater was sent to a 5,000-ton storage tank via pipeline.

Collected contaminated groundwater was sent to a dedicated distillation unit at the plant for recovery. At the end of the first year of operation, the acrylonitrile concentrations in the extracted water decreased significantly below recovery efficiency level of the distillation unit. Thence, the extracted water was diverted to the facility's wastewater treatment unit until the end of the extraction program. Groundwater extraction and treatment operation continued for 4 years, until the end of 2004. A total of 53,000 m^3 of groundwater was extracted and treated. The quantity of pumped and treated groundwater is presented in Figure 5.

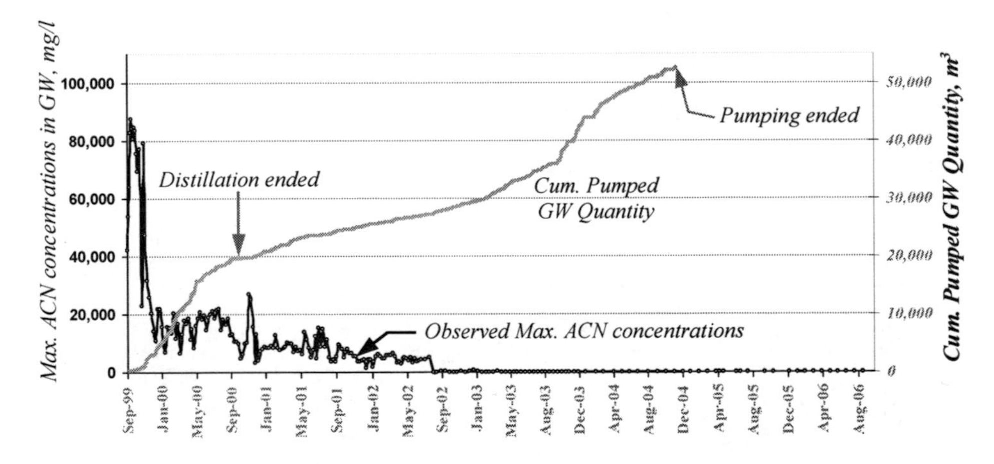

Figure 5. Pumped and treated groundwater (GW) quantities and ACN concentrations

2.3. ON-SITE CONTAINMENT OF CONTAMINATED GROUNDWATER

The continuous groundwater extraction at the wells and drainage trenches further enhanced the natural onsite containment of the contaminated groundwater by lowering the water levels underneath the tank farm area as shown in the schematic block diagram and a recorded groundwater level graph are presented in Figure 6 (a) and (b).

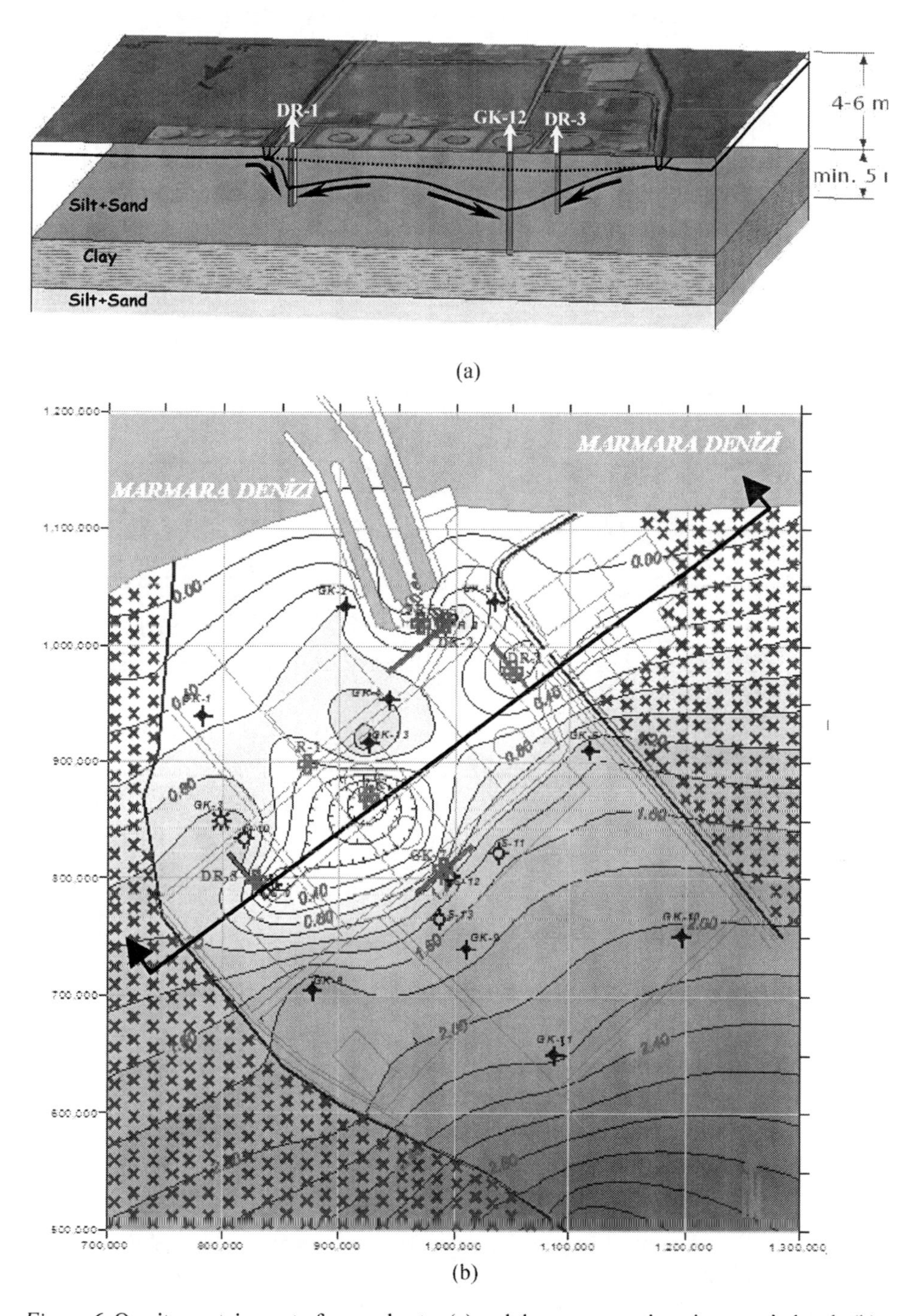

Figure 6. On-site containment of groundwater (a) and the representative piezometric levels (b)

2.4. GROUNDWATER QUALITY MONITORING RESULTS

During this seven year long project, 243 rounds of groundwater samples, collected from 24 locations, were analyzed for acrylonitrile. Piezeometric level and ACN distribution maps for each sampling round were used in optimization of the groundwater extraction efficiency for removal of the residual contamination in the aquifer. A sample of the ACN distribution map and relevant recorded parameters are presented in Figure 7.

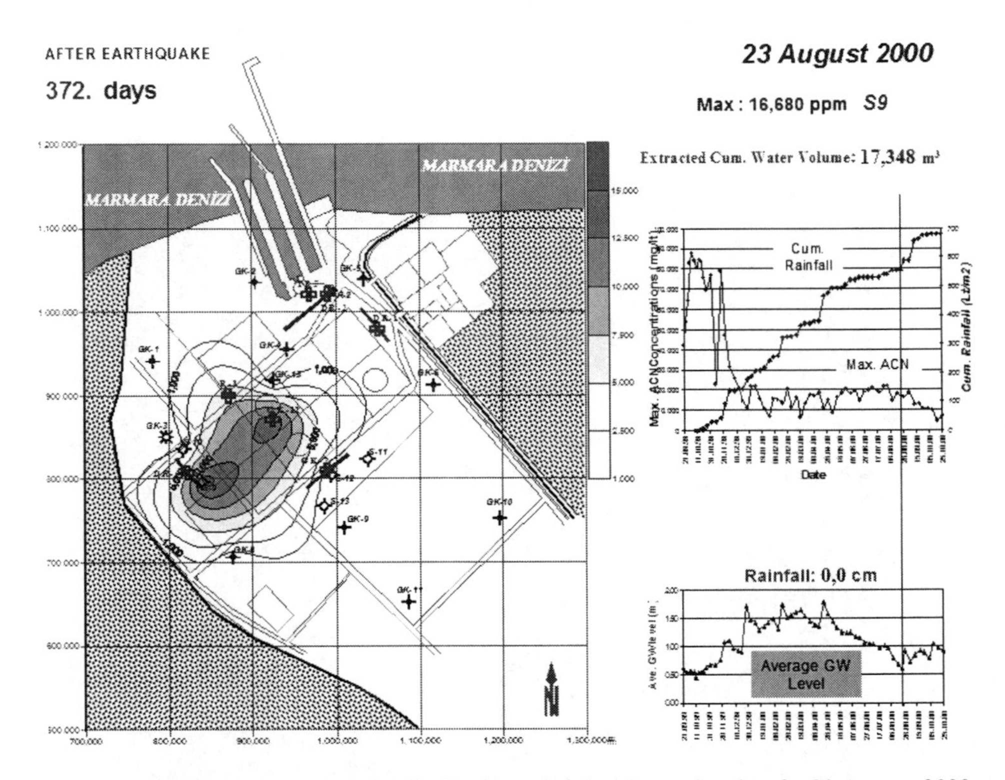

Figure 7. Graphical records of ACN distribution, rainfall and pumping data for 23 August, 2000

3. Contaminant Removal - Model

At the onset of the groundwater extraction activities, the following symptomatic relationship between the piezometric level, ACN concentrations and rainfall was observed at the extraction wells:

- The ACN concentrations at the extraction well points showed a systematic variation following restart of pumping after pump malfunctions. Such observations were tracked and it was concluded that the ACN concentrations showed an increasing trend as the piezometric levels at the extraction

wells were lowered during the groundwater extraction. As pumping was halted for a couple of days, the ACN concentrations decreased.
- ACN concentrations showed a general decreasing trend as the piemetric levels in all monitoring locations rose after intermittent rainfall events.

ACN is a highly polar organic chemical and has a "very low to negligable" adsorbtion potential to clay and other minerals in soils. The observations above led us to a hypothesis that ACN present in the unsaturated soil between the surface and the water table was scrubbed into the shallow aquifer by the percolating rainfall and then effectively dragged into the extraction well by the increased hydraulic gradient of the extraction well drawdown cone. This concept was applied to the groundwater extraction scheme with the programmed halt-and-go schedules in order to optimise the ACN extraction efficiency.

A seven-year record of maximum ACN concentrations, average groundwater level, annual cumulative rainfall and cumulative groundwater extraction volume are presented in Figure 8.

A conceptual model for the experimentally-verified soil washing/water extraction process is presented in Figure 9.

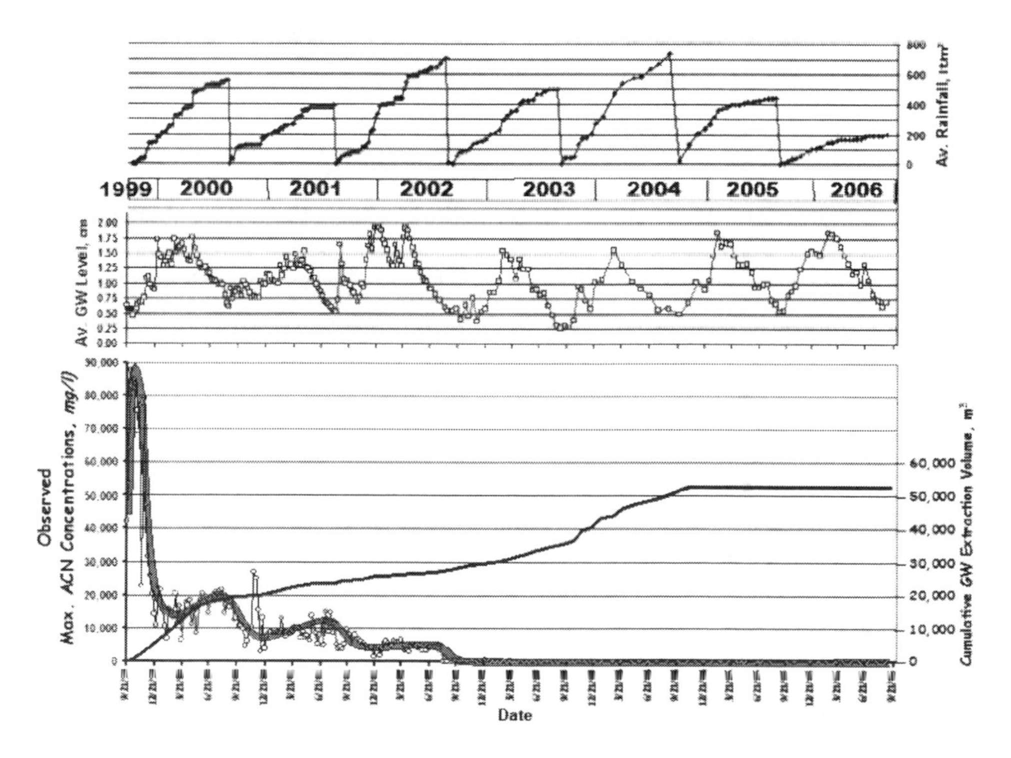

Figure 8. Seven-year record of the maximum ACN concentrations, average groundwater level, annual cumulative rainfall and cumulative groundwater extraction volume

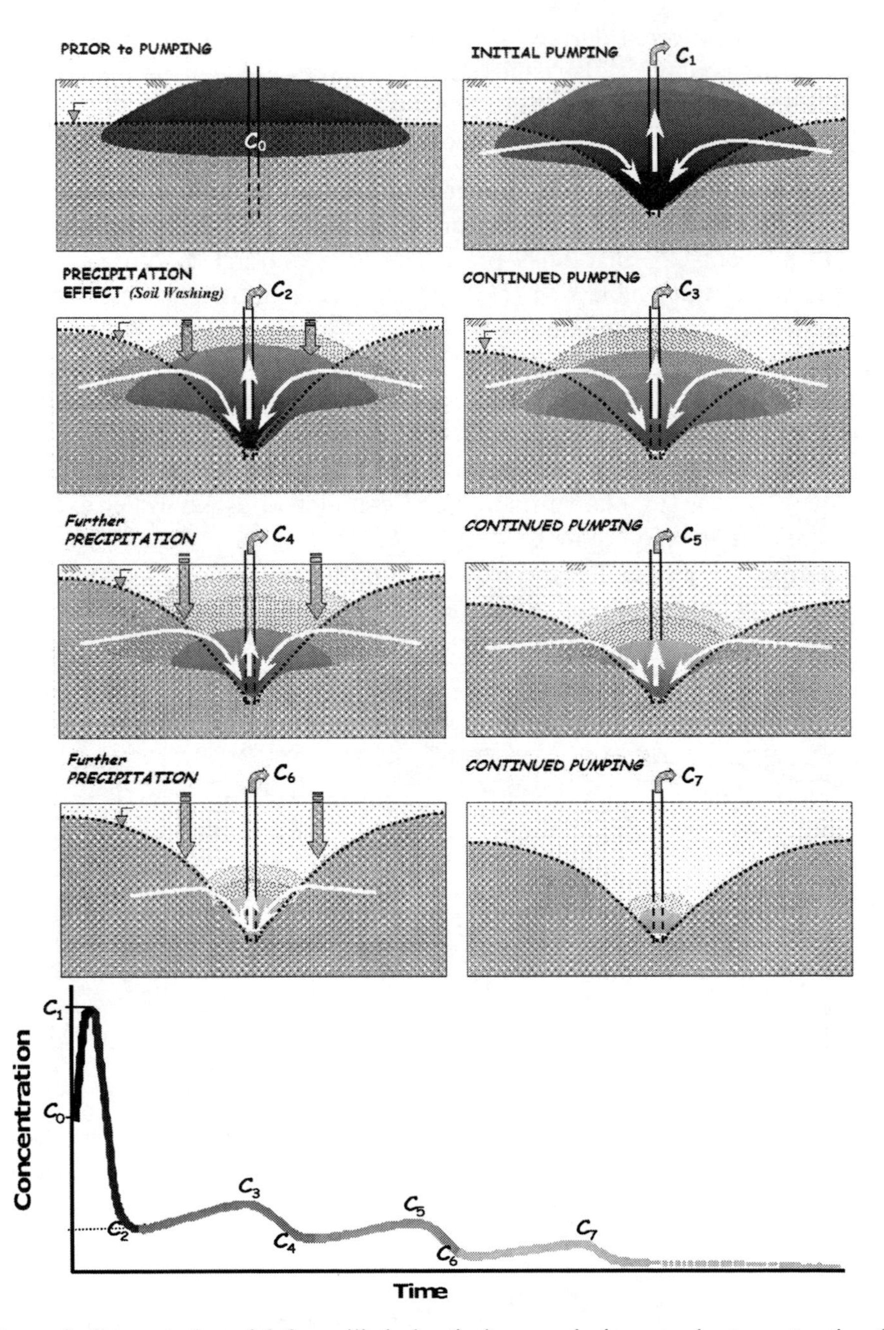

Figure 9. Conceptual model for spilled chemical removal via groundwater extraction from granular non-adsorbing medium

4. Conclusions

Earthquakes are natural events that have been causing disaster level disruptions in all aspects of human life. Disaster preparedness is a key factor in minimizing the impacts of such events on the communities. The case history presented in this paper reveals that fatalities and fires at major industrial facilities can be prevented with properly taken process safety measures, even under the worst post-disaster circumstances and major chemical spills can be remediated effectively.

The case history presented in this paper reveals also that a good understanding of the local geology, hyrogeological setting and implementation of an effective pump/treat scheme, optimized with continuous monitoring of some basic parameters (water quality, piezometric levels and precipitation) play a key role in the success of on-site containment of chemical spills and remediation of shallow aquifers.

Acknowledgement

The author extends his sincere thanks to the management of the AKSA Akrilik Kimya AS of Turkey for their conscientious efforts for immediate initiation of the groundwater quality monitoring and remediation program. Special appreciation is also extended to the memory of late Mr. S. Ergin and to Mr. M. Yilmaz, past and current general managers of the company for their dedication to this rigorous monitoring and extraction efforts that continued for seven years. Special appreciation goes to the Turkish Chemical Manufacturers Association for its continued technical support for the project under the Responsible Care® initiative of the world chemical industry. This paper was presented at the NATO Advanced Research Workshop in Sinaia, Romania on 10 October 2006 as an invited lecture.

References

Steinberg, L.J., Cruz, A.M., Vardar-Sukan, F. and Ersoz, Y. (2001), "Hazardous Materials Releases during the August 17, 1999 Earthquake in Turkey", Proc. World Water and Environmental Resources Congress, ASCE, May 20–24, Orlando, Florida, USA, 10.1061/40569(2001)445.

ICCA-UNEP (2002a), "Chemicals-Industry as a partner for sustainable development", Report contribution to the World Summit on Sustainable Development. The International Council of Chemical Associations and UNEP, ISBN: 92-807-2197-8, p. 48.

Sengor, S.S. (2002b), "Modeling Contaminant Transport and Remediation at the AKSA Acrylonitrile Spill Site", M.Sc. Thesis, Middle East Technical University, Ankara, Turkey, 103p.

HOW TO REMEDIATE POLLUTION WITH MERCURY AND HEXACHLORO-CYCLOHEXANE RESIDUES IN A CHEMICAL PLANT

SVETOMIR HADZI JORDANOV*
Faculty of Technology & Metallurgy, University UKIM Skopje, R. Macedonia

Keywords: Groundwater, Contamination, Metals, Remediation

1. Summary

Pollution of soil and underground water in a major chemical plant situated in Macedonia's capital Skopje started some 30 years ago. Brine electrolysis plant was inevitably a source of pollution due to the (i) use of enormous quantities of mercury as a cathode in the electrolysis cells and (ii) a long and complex scheme of mercury circulation through the unit segments. Spilling of mercury is still present in soil and groundwater in concentrations that significantly exceed the legal allowances, no matter that plant' operation was ceased 15 years ago.

Synthesis of hexachloro-cyclohexane (HCH), whose γ-isomer was the active component of insecticide 'lindane', was also a source of significant soil and groundwater pollution. 86 – 88% of each HCH-batch was of non-γ-isomers, so that some 30.000 - 35.000 tons of HCH-isomers did accumulate during the 14 years of HCH-unit operation. The by-products were temporary disposed in a concrete pool constructed specially for the purpose and covered with non-permeable cover and top soil layer. As usually happens, nothing is as long-lasting as the temporary attempts are, and today, some 30 years latter, the HCH residues are still in their concrete pool. Unfortunately, the pool's content became accessible to rainfalls and the pool bottom started to leak, so that the soil and groundwater show results of heavy contamination with chlorinated hydrocarbons and products of HCH-decay.

* Svetomir Hadzi Jordanov, Faculty of Technology & Metallurgy, University UKIM Skopje, R. Macedo, shj@tmf.ukim.edu.mk

M.D. Annable et al. (eds.), Methods and Techniques for Cleaning-up Contaminated Sites, 103–120.

In the meantime, with the rise of environmental concern, the problem was recognized, precisely identified and documented. Some proposals for its solution were elaborated, but the main objective remains untouched - how to provide funding for such an expensive treatment in a country that faces a lot of serious economic problems.

This paper deals with a number of details aimed at helping remediation of above described pollution. The aim is, if there is no possibility for urgent solving of the problem, just to keep it present on the agenda, and to stress it permanently until the solution is applied.

2. Background

HCH (1,2,3,4,5,6-hexachlorocyclohexane, $C_6H_6Cl_6$) also known as Lindane, is a chlorinated organic pesticide, colorless, crystalline solid with faint or no smell. After the World War II it was one of the two insecticides with the largest volume of production (next to DDT) and widely used in agriculture (Van Nostrand's, 1995).

Lindane is more toxic to insects than DDT and it bio-accumulates rapidly in organisms. It has a half-life of 2.3 to 13 days in air, 30 to 300 days in water, 50 days in sediments and two years in soil. It is stable to light, high temperatures and acid but it can be hydrolyzed at high pH. Lindane degrades very slowly by microbial action. It is more water-soluble and volatile than other chlorinated organic chemicals.

Considering the health impact, lindane is considered to be a carcinogen and has been associated with liver cancer. It is especially toxic for children who have large body surface as an absorption field. It has been used on the skin against scabies and on the head against lice. It could be absorbed from the air through respiration in the vicinity of factories and may be ingested through contaminated vegetables, meat or water (the most common human exposure pathway to lindane is through food). Lindane may also pass through breast-feeding.

Medical studies have shown that there is evidence of antiandriogenic effects (cryptorchidism, atrophy of the testes, lower androgen levels in the blood), estrogenic effects (breast carcinoma) etc. Malformation in newborns could also be caused by exposure to lindane. Nevertheless, the elimination of lindane from the body is fairly rapid once its use has been discontinued.

The production of HCH was stopped worldwide after it was recognized how harmful is the impact of organic insecticides upon the total environment – man, animals and fowl, soil inhabiting microflora and microfauna and all forms of aquatic life. The effects of these insecticides upon population of beneficial insects were as well critically assessed.

In a period of 1964-77, this OHIS Plant did formulate the insecticide *Lindane* (active compound + carrier). The active compound, gamma-hexachloro-cyclohexane, was produced by photo-chlorination of benzene, C_6H_6. The produced mixture contains also other HCH-isomers, i.e. *alpha*, *beta* and *delta*. Being poor insecticides, these ballast isomers were treated as a waste and some 30,000 - 35,000 t were 'temporary' discharged in a dumpsite (Jordanov, 2005).

A concrete pool was erected, filled with the ballast HCH-isomers and covered with a layer of soil. The top of the pool was not properly protected against penetration of rainfall, and the bottom was not properly protected against leakage. The pool is approximately 100 meters long, 50 meters wide and several meters high (See Fig. 1) and is situated near the main river Vardar. It was constructed without a drainage system for collecting percolating liquids and without a cover to prevent leaching. If barrels were used for storage, it is likely that they have corroded. The area around the basin smelled of chlorinated compounds.

Figure 1. General view of the lindane dump site

3. Brief History of Lindane Production in OHIS-SKOPJE

HCH was produced by photosynthesis of a mixture of benzene and chlorine. The product of synthesis contains 4 or 5 isomers of HCH, named with the first letters of Greek alphabet. Among them only the γ–isomer was used for formulation of the insecticide, while the rest of isomers were only a ballast byproduct. Unfortunately the purchased technology (State-of-the-Art of the 1950s) had a poor yield – as much as 86 to 88 percent (m/m) of the produced HCH were of inactive isomers. The gamma isomer was separated by extracting in methanol. With further concentration process, a 99.9% concentrated gamma isomer was produced. Methanol was completely recycled and was not incorporated in the final product.

After extraction of γ–isomer, the rest of the product mixture was either decomposed to trichlorobenzene and hydrochloric acid, or simply – dumped out.

Lindane production was stopped due to problems with placement, but also and due to environmental concerns. The following was stressed out as main reason for stopping the lindane production (Jordanov, 2005):

1. Lack of appropriate procedure for valorisation of inactive isomers up to products with higher commercial value. As much as 88% of the synthesized HCH are inactive isomers that in the applied technology were concerned as waste, no mater that in some other ones they are concerned as valuable raw-material for production of insecticides and fungicides.
2. The Lindane unit capacity was small to be competitive, but still too large for domestic consumption only. Appearance of more effective and less toxic lindane substitutes, as well as imposing of new bans of its application further limited its consumption.
3. The burden of vast quantities of inactive isomers created not only environmental problems, but also created financial ones, due to the transportation and storage of waste.

For similar reasons also some other HCH production plants in Europe were closed.

4. Properties of HCH

Some of physical properties of HCH are given bellow (Kirk-Othmer, 1993):

Alpha and *beta*-isomers

- Molecule mass: 290.9
- Density: 1.95 g/cm^3
- Melting point: 183°C
- Appearance: white powder mater
- Odour: unpleasant

- Soluble in: acetone, methanol and other organics
- Non-soluble in: water
- Heat of combustion: 11.3 kJ/kg
- Volumetric mass of dry waste: 0.608 g/cm^3 (non-compacted) 0.888 g/cm^3 (compacted)
- Average abundance in the isomer's mixture:

alpha isomer	84.75%
gamma + etha	0.55%
beta isomer	12.27%
water content	0.5%
methanol content	0.5%

***delta*-past**

- Appearance: dark brown past
- Smelt: unpleasant (to HCH)
- Soluble in: acetone, methanol, benzene
- Non-soluble in: water
- Specific mass at 20°C: 0.950 g/cm^3
- Flash point: inflammable
- Explosiveness: non-explosive
- Liquefies at: 50 - 60°C
- Average content of isomers:

alpha	22 - 26%
beta	5 - 7%
gamma	16 - 19%
delta	38 - 50%

- Other components: chlorinated benzene and fatty mater of unknown composition
- Isomer's toxicityLD50 (rats):

alpha	500 mg/kg
beta	6000 mg/kg
gamma	125 mg/kg
delta	1000 mg/kg
etha	1000 mg/kg

5. Experimental

Recently the state of the lindane dump site was studied as described elsewhere (UNEP, 2001), (Jordanov, 2003), (MOL, 2002). Two holes were drilled under the most suspected locations down to a dept of 12 meters (See Fig. 2). The

Figure 2. Lindane dump site piezometer

drilling cores were used as samples for soil chemical analysis. Two piezometers were inserted into the holes and samples of underground water were pumped out for the same purpose. Water from a waste water treatment unit as well from Vardar River downstream the OHIS plant, as well as sediments from the bottom of a previously used wastewater canal were sampled. All samples were properly prepared and subjected to chemical analysis by a specialized service laboratory (MOL, 2002).

5.1. *Extraction of the Water Samples*

The water sample (100 ml) has been extracted by 3 × 1 ml of pentane. The upper pentane layer was isolated and dried by anhydrous Na_2SO_4. All pentane extracts were analyzed by Gas Chromatography.

5.2. *Extraction of Sediment Samples*

A sample of earth was three times extracted by a mixture of hexane and acetone. The extract was concentrated to 1 ml, and then analyzed by Gas Chromatography.

Gas Chromatograph Hewlett-Packard HP 5890 Series II was employed.

5.3. *Mercury*

Mercury was determined by the cold steam method on a MERCURY ANALYZER 254.

6. Results

In tables 1 - 4 concentration of *Lindane*, trichloroethylene, total chlorinated organics (as *Lindane*), other POPs and mercury measured in the year 2002 are shown (MOL, 2002).

TABLE 1. Concentration of pollutants (μg/L) in ground water

Parameter	Lindane plant	Electrolysis plant
Lindane	1.15	/
Trichloroethylene	/	2370
Total chlorinated organics (as Lindane)	1017	/
Mercury	2.28	2.78

From Table 1 it is seen that lindane and its decay products as well as mercury are present in groundwater in concentration that exceed the legal allowances (0,01 μg/L for lindane and 0.2 μg/L for mercury).

TABLE 2. Concentration of pollutants in soil

Parameter	Lindane plant		Electrolysis plant	
	Depth, m	μg/kg	Depth, m	μg/kg
Lindane	3-4	0.57	/	/
	7-8	<0.50		
	11-12	<0.50		
Trichloroethylene	/	/	4-5	0.22
			9-10	0.06
Total chlorinated organics (as Lindane)	3-4	7192	4-5	127.6
	7-8	9.69	9-10	42.7
	11-12	32.8		
Mercury	0-12	<0.10	0-7	<0.10
			8-9	0.17
			10 11	0.10
			11-12	0.58

From Table 2 it is seen that concentrations of lindane, trichloroethylene and total chlorinated organics in soil vary significantly with the depth. The randomness of the variation indicates the need for application of statistical treatment of results. Concentration of mercury in soil under the former Electrolysis plant rises regularly with the depth, as a result of mercury's high specific mass. It is possible that bellow 12 m mercury concentration continues to increase.

TABLE 3. Concentration of pollutants (µg/L) in wastewater before the inlet and in Vardar River immediately downstream the OHIS plant

Parameter	Wastewater before inlet in river	River water downstream the plant
Trichloroethylene	23.40	0.25
Chloroform	6.9	13.40
1,2 Dichloroethane	10.70	20.70
Tetrachloroethylene	0.434	0.020
1,2 Dichlorobenzene	0.9	<0.5
Benzene, Xilene, Toluene, Ethylbenzene (each)	<1.0	<1.0
PCBs	<1.0	<1.0
Mercury	0.11	0.72

From Table 3 the effect of dilution of plant's wastewater is seen. Concentrations of tri- and tetra-chloroethylene and 1,2 dichlorobenzene fall significantly after inlet of wastewater into the river. Increase of the concentrations of chloroform and 1,2 dichloroethane downstream the plant could indicate that these products of lindane decay enter the river not through the wastewater stream, but through the groundwater. Concentration of mercury is 6.5 times higher downstream the plant and this increase is, again, caused not by inlet of visible wastewater but by groundwater polluted with mercury.

TABLE 4. Concentration of pollutants in soil from the bottom of previously used wastewater canal

Depth, m	Lindane µg/kg	Mercury µg/kg	Depth, m	Lindane µg/kg	Mercury µg/kg
0.0 – 0.5	53.9		2.5 – 3.0	0.24	
0.5 – 1.0	42.7		3.0 – 3.5	3.50	
1.0 – 1.5	11.0	<0.10	3.5 – 4.0	4.75	<0.10
1.5 – 2.0	4.94		4.0 – 4.5	4.94	
2.0 – 2.5	29.5		4.5 – 5.0	1.22	

From Table 4 it is seen that as a rule the concentration of lindane in the soil of the former wastewater canal decreases with depth up to 5 m, while the concentration of mercury is practically the same (<0.10 μg/kg). Mercury, if present probably did penetrate deeper in the soil.

More detailed analyses were made in the year 2005, as shown in Tables 5 – 19 (MOEPP, 2005).

6.1. CHARACTERISTICS OF GROUNDWATER

6.1.1. *Groundwater near the Pesticide Dump Site*

In Tables 5 - 8 concentrations of pollutants in groundwater sampled in the year 2005 are shown.

TABLE 5. Chemical analysis (μg/L) of heavy metals in groundwater near pesticide dump site

Parameter	Max. allowed	Measured	Method
pH	6,5-8,5	7.70	ISO 10523
Conductivity, μS/cm	-	1291.0	ISO 7888
Lead, Pb	10	17,519	ISO 11885/1996
Mercury, Hg	0,2	0,11	ISO 11885/1996
Chromium, Cr	50	14,49	ISO 11885/1996

From Table 5 it is seen that lead is the only heavy metal that exceeds the allowed concentration in groundwater near pesticide dump site.

TABLE 6. Chemical analysis (μg/L) of pesticides in groundwater near pesticide dump site (Method US EPA 8081A/8270)

Parameter	Max. allowed	Measured
α-HCH	0.01	2.40
γ-HCH	0.01	0.38
β-HCH	0.01	3.20
δ-HCH	0.01	0.08
Aldrin	0.003	0.08
DDE	0.001	0.00167
(+)-cis hepta- chloroepoxide	0.001	0.2
2,4'DDD	0.001	0.004

From Table 6 it is seen that concentration of all measured pesticides in groundwater near pesticide dump site exceeds the legal allowances. The excess of β-HCH is 320 times over the limit.

TABLE 7. Chemical analysis (μg/L) of extractable organochalogens (EOX) in groundwater near pesticide dump site (Method US EPA 524.2)

Parameter	Max. allowed	Measured
$CHCl_3$	2	0.543
CCl_4	2	0.235
C_2HCl_3	3	0.164
$CHCl_2Br$	2	0.48
C_2Cl_4	2	-
$CHBr_3$	2	18.39

From Table 7 it is seen that methyl-tribromide $CHBr_3$ is present in groundwater in concentration almost 10 times higher than the legal allowance. All other extractable organochalogens are present in allowed concentrations.

TABLE 8. Chemical analysis (μg/L) of polycyclic aromatic hydrocarbons (PAH) in groundwater near pesticide dump site (Method US EPA 610/625)

Parameter	Maximum allowed	Measured	Parameter	Maximum allowed	Measured
naphthalene	1	0.030	krizen	0.01	< D. L.
fenantrene	5	0.009	benz(b)fluorantrene	0.01	< D. L.
acenaphtene	5	< D. L.	benz(k)fluorantrene	0.01	< D. L.
antracene	5	< D. L.	benz(a)pyrene	0.01	< D. L.
fluorantrene	5	0.008	indeno(1,2,3,cd)pyrene	0.01	< D. L.
pyrene	0.01	0.015	dibenz(a,h)antracene	0.01	< D. L.
benz(a)antracene	0.01	< D. L.	benzo(g,h,i)perylene	0.01	< D. L.

From Table 8 it is seen that pyrene is the only PAH present in groundwater in slight excess to the legal allowance.

6.1.2. *Groundwater near the Former Electrolysis Plant*

In Tables 9 - 12 concentrations of pollutants in groundwater sampled in the year 2005 are shown.

TABLE 9. Chemical analysis (μg/L) of heavy metals in groundwater near former electrolysis plant

Parameter	Max. allowed	Measured	Method
pH	6.5-8.5	8.18	ISO 10523
Conductivity, μS/cm	-	1431	ISO 7888
Lead, Pb	10	4,964	ISO 11885/1996
Mercury, Hg	0.2	1.10	ISO 11885/1996
Chromium, Cr	50	28.42	ISO 11885/1996

From Table 9 it is seen that mercury is present in groundwater near former electrolysis plant in concentration 5.5 times higher than the legal allowance.

TABLE 10. Chemical analysis (μg/L) of pesticides in groundwater near former electrolysis plant (Method US EPA 8081A/8270)

Parameter	Max. allowed	Measured
α-HCH	0.01	0.239
γ-HCH	0.01	0.0252
β-HCH	0.01	0.282
δ-HCH	0.01	-
Aldrin	0.003	0.3
DDE	0.001	0.0225
(+)-cis hepta-chloroepoxide	0.001	-
2,4'DDD	0.001	-

From Table 10 it is seen that all analyzed pesticides are present in groundwater near the former electrolysis plant in concentration higher than the legal ones. Aldrin exceeds 100 times the legal allowance. β-HCH exceeds the allowance for 28 times, as compared with 320 times in groundwater near the pesticide dump site. Distance between the two piezometers is about 500 m and the measured gradient in β-HCH concentration indicates that there is still ongoing leakage of pesticides from the pesticide dump site.

TABLE 11. Chemical analysis (μg/L) of extractable organochalogens (EOX) in groundwater near former electrolysis plant (Method US EPA 524.2)

Parameter	Max. allowed	Measured
$CHCl_3$	2	0.758
CCl_4	2	0.797
C_2HCl_3	3	104.9
$CHCl_2Br$	2	0.5
C_2Cl_4	2	132.4
$CHBr_3$	2	-

From Table 11 it is seen that tri- and tetra- chloroethylene are present in groundwater in excess of 35 and 66 times, respectively. All other extractable organochalogens are present in allowed concentrations.

TABLE 12. Chemical analysis (μg/L) of polycyclic aromatic hydrocarbons (PAH) in groundwater near former electrolysis plant (Method US EPA 610/625)

Parameter	Maximum allowed	Measured	Parameter	Maximum allowed	Measured
naphthalene	1	0.045	krizen	0.01	< D.L.
fenantrene	5	0.008	benz(b)fluorantrene	0.01	< D.L.
acenaphtene	5	< D.L.	benz(k)fluorantrene	0.01	< D.L.
antracene	5	< D.L.	benz(a)pyrene	0.01	< D.L.
fluorantrene	5	0.005	indeno(1,2,3,cd) pyrene	0.01	< D.L.
pyrene	0.01	0.010	dibenz(a,h)antracene	0.01	< D.L.
benz(a)antracene	0.01	< D.L.	benzo(g,h,i)perylene	0.01	< D.L.

From Table 12 it is seen that in groundwater near the former electrolysis plant there are no PAH in excess over the allowances.

6.2. CHARACTERISTICS OF SOIL

6.2.1. *Soil near the Pesticide Dump Site*

In Table13 physical analysis and in Tables 14 - 16 concentrations of pollutants in soils sampled in the year 2005 are shown.

TABLE 13. Physical analysis of soil near the pesticide dump site

		Fractions, %					
Piezometer	Humus	Sand		Dust	Clay	Clay + dust	Sand
core, m	%	0.2-2	0.02-0.2	0.002-0.02	< 0.002	< 0.02	0.02-2
0.3-6.7	0.17	12.2	57.8	17.6	12.4	30.0	70.0

TABLE 14. Chemical analysis (mg/kg) of soil near the pesticide dump site (Method ISO 11885/1996)

Parameter	Max. allowed	Measured
Lead, Pb	85	12.44
Mercury, Hg	0.3	0.12
Chromium, Cr	100	84.76

From Table 14 it is seen that in the soil near the pesticide dump site there is no heavy metal in excess.

TABLE 15. Chemical analysis (mg/kg) of pesticides in soil near the pesticide dump site (Method US EPA 8081A/8270)

Parameter	Max. allowed*	Measured
α-HCH	0.003	1.26
γ-HCH	0.00005	0.26
β-HCH	0.009	2.12
δ-HCH	-	0.11
Aldrin	0.00006	1.83
Dieldrin	0.005	0.01
DDE	0.01	0.01

*Dutch standards

From Table 15 it is seen that all but DDE pesticides are present in the soil near the pesticide dump site in high excess, as compared to Dutch standards. Aldrin exceeds the allowance for 30,000 times and γ-HCH for 5,200 times! This is a proof that the soil near the pesticide dump site is heavy contaminated.

TABLE 16. Chemical analysis (mg/kg) of polycyclic aromatic hydrocarbons (PAH) in soil near pesticide dump site (Method US EPA 8275A)

Parameter	Maximum allowed	Measured	Parameter	Max. allowed	Measured
Total PAH	1	0.153	krizen	-	0.012
naphthalene	-	0.028	benz(b)fluorantrene	-	< 0.01
fenantrene	-	0.020	benz(k)fluorantrene	-	< 0.01
antracene	-	< 0.01	benz(a)pyrene	-	< 0.01
fluorantrene	-	0.039	indeno(1,2,3,cd)pyrene	-	< 0.02
pyrene	-	0.044	dibenz(a,h)antracene	-	< 0.02
benz(a)antracene	-	0.010	benzo(g,h,i)perylene	-	< 0.02

From Table 16 it is seen that PAH are not present in excess in the soil near the pesticide dump site.

6.2.2. *Soil near the Former Electrolysis Plant*

In Tables 17 - 19 concentrations of pollutants in soils sampled in the year 2005 are shown.

TABLE 17. Chemical analysis (mg/kg) of heavy metals in soil near the former electrolysis plant (Method ISO 11885/1996)

Parameter	Max. allowed	Measured
Lead, Pb	85	14.72
Mercury, Hg	0.3	7.00
Chromium, Cr	100	75.2

From Table 17 it is seen that mercury is present in the soil near the former electrolysis plan in excess of over 20 times.

From Table 18 it is seen that pesticides are present in the soil near the former electrolysis plant on excess of 1,000 times (aldrin) and 100 times (α-HCH and γ-HCH). These figures should be compared with 30,000 times and 5,200 times for the pesticides excess near the pesticide dump site (Table 15). Again a gradient in pesticide's concentrations is obvious.

TABLE 18. Chemical analysis (mg/kg) of pesticides in soil near to former electrolysis plant (Method US EPA 8081A/8270)

Parameter	Max. allowed	Measured
α-HCH	0.003	0.352
γ-HCH	0.00005	0.0043
β-HCH	0.009	0.27
δ-HCH	-	-
Aldrin	0.00006	0.06
Dieldrin	0.005	-
DDE	0.01	0.018

TABLE 19. Chemical analysis (mg/kg) of polycyclic aromatic hydrocarbons (PAH) in soil near the former electrolysis plant (Method US EPA 8275A)

Parameter	Maximum allowed	Measured	Parameter	Max. allowed	Measured
Total PAH	1	0.155	krizen	-	0.013
naphthalene	-	0.052	benz(b)fluorantrene	-	< 0.01
fenantrene	-	0.023	benz(k)fluorantrene	-	< 0.01
antracene	-	< 0.01	benz(a)pyrene	-	< 0.01
fluorantrene	-	0.018	indeno(1,2,3,cd) pyrene	-	< 0.02
pyrene	-	0.034	dibenz(a,h)antracene	-	< 0.02
benz(a)antracene	-	0.015	benzo(g,h,i)perylene	-	< 0.02

From Table 19 it is seen that PAH are not present in excess in the soil near the former electrolysis plant.

7. Conclusions

Results presented in this study could be summarized in the following conclusions.

1. All groundwater is heavily contaminated with lindane and decay products. The most contaminated is the aquifer under the pesticide dump site where pesticides are present in excess of the legal allowances up to 320 times! Some 500 m aside from this excess falls approximately 10 times. This gradient in concentration is caused by constant leakage of pesticides from

the dump site and the radial spreading (and dilution) around the source of contamination. This contamination reaches the nearby Vardar River where increased concentrations of lindane decay products were measured. Sediments in the former wastewater canal also contain increased levels of pesticides.

2. Groundwater is also contaminated with mercury, especially under the former electrolysis plant. Contaminations with mercury have reached the Vardar River.
3. Soil up to 12 m is also contaminated with lindane and decay products and mercury. While lindane and decay-products are concentrated in the upper layers, mercury penetrates deep in the soil. It could be expected that bellow 12 m its concentration is even higher.
4. Surface soils are more heavily contaminated with pesticides than other tested media. Near the pesticides dump site excess of 30.000 times of aldrin was measured.

8. How to Remediate?

The goal of this study, as stressed in its title, is how to remediate the pollution with mercury and HCH residues. This goal could possibly be achieved in the following steps:

1. Determination of the extent of contaminated area

The limits of both contaminated soil and aquifer under the existing HCH dump site and former electrolysis plant are to be determined as precisely as possible. In order to do that a network of testing points is to be designed according to the principles of hydrogeology and the existing local conditions. In each chosen testing point a hole should be drilled up to the depth that reaches the lowest level of soil or water contamination. A piezometer is to be inserted in each hole, so that long-term water sampling (monitoring of underground water contamination) could be done on a regular basis. Cores from drilling cylinders should be used as samples for analyzing the soil contamination.

2. Determination of the level of contamination

The level of contamination of both soil and underground water should be determined by performing systematic analysis of the content of HCH and products of its decay, as well as mercury in soil and water samples.

Water sampling should be repeated many-fold, so that data acquisition for statistical analysis or for following the variations of the contamination with the seasons of the year could be possible.

3. Remediation of the contaminated soils

Remediation of the contaminated soil is the most serious problem. It could be done by striping off the heavily contaminated (surface) layers and by chemical treatment of the less contaminated ones. The later treatment is rather dubious. Both are extremely expensive.

The real costs of this operation will be known only after completion of previous steps of the integral Project.

4. Remediation of contaminated aquifers

Remediation of the contaminated aquifer is planned to be done by pumping out vast quantities of water with such an intensity that depression funnel will be formed around the pumping point. The aim is to direct the contaminated water towards the depression center and to repeat the rinsing of underground as long as proper cleaning is achieved. The pumped water should be purified by passing through an adsorbing media, e.g. activated carbon, bentonite or similar, and then fed back to the periphery of the underground aquifer.

5. Excavation of HCH waste dump and incineration of waste

This final phase of the waste remediation should cover a number of aspects as, e.g.:

- Preparing a remedial plan, including a detailed work plan and a safety plan,
- Excavation of chemical waste and contaminated soil underneath the dumps,
- Transportation the excavated waste and soil to the incineration plant,
- Incinerating the waste and soil,
- Backfilling the excavation pit with clean soil or treated soil from the incineration process,
- Reusing the area, for example, for industrial development,
- Disposing of waste water treatment residues and incinerating residues.

There is no need to repeat that all these steps are extremely costly. Neither the company nor the state will be able to solve it in near future.

References

Jordanov S. H., Dika D. and Paunovic P. (2003) Mercury Cell Brine Electrolysis and Lindane Plant as Sources of Persistent Water and Soil Contamination, *Chemistry and the Environment, II Regional Symposium,* Serbia & Monte Negro, Proceedings, p. 207

Jordanov, S. H. and Tanevski J. (2005) Identification of pollutants and their sources in the plants of OHIS Skopje, Environmental study, Skopje 2005

Van Nostrand's Scientific Encyclopedia, 8th ed. (1995), A-I, Van Nostrands Reinolds, N.Y. et al., p. 1723

Kirk-Othmer Encyclopedia of Chemical Technology, S.H.Safe, vol. 6, John Wiley & Sons, N.Y., 4th ed., 1993, pp 135-139

UNEP Feasibility Study for Urgent Risk Reduction Measures at Hot Spots in Macedonia (2001) – only for Waste on Lindane

MOL – Belgrade (2002), Test Report on the characterization of pollution in OHIS Skopje, Belgrade

MOEPP (2005), Study on the level of contamination in the Industrial Hot Spots, Skopje, 2005

DECISION SUPPORT SYSTEM FOR EVALUATION OF TREATMENT TRAIN FOR REMOVAL OF MICROPOLLUTANTS

PETR HLAVINEK, JIRI KUBIK
Brno University of Technology, Faculty of Civil Engineering, Institute of Municipal water Management, Zizkova 17, 602 00 Brno, Czech Republic

Abstract. European Water Framework Directive (WDF) aims to achieve and maintain European water bodies in “good status”. Micropollutants such as pesticides, surfactants, preservatives, solvents, fragrances, flavors and pharmaceuticals as well as endocrine disruptors are of rising concern in the urban water cycle. With the growing number of efficient and innovative treatment processes, evaluation and selection of the appropriate configuration of treatment processes is a challenging task. A process simulation model has been developed to allow a range of design possibilities for removal of micropollutants to be evaluated. The model includes a computational module for wastewater treatment trains, a database and an optimization engine.

Keywords: Micropollutants, Decision support, Simulation, Treatment trains

1. Micropollutants

Xenobiotics such as pesticides, surfactants, preservatives, solvents, fragrances, flavours, and pharmaceuticals as well as endocrine disrupters are of rising concern in the urban water cycle. There are > 100,000 xenobiotics on the market in the European Union. Approximately 30,000 of these are “everyday” chemicals i.e. estimated to be used in volumes over 1 ton each year. It has been estimated that 70,000 xenobiotics may potentially be hazardous for humans and/or ecosystems. In order to assess the role of the xenobiotics, information is needed with respect to the sources, flow paths, fate (transport, treatment, natural attenuation) and impact on both humans, livestock and the environment. Furthermore, it is necessary to have suitable tools like chemical analytical methods or eco-toxicological test methods for collecting the information that is needed and assessing the potential risk.

M.D. Annable et al. (eds.), Methods and Techniques for Cleaning-up Contaminated Sites, 121–127.

The integrated nature of urban water systems and the ability of xenobiotics to spread across structural boundaries and into the environment where ecological systems and humans are exposed calls for an intersectorial and multi-disciplinary approach to problem awareness and solution.

Innovative approaches are needed to prevent xenobiotics from being discharged into surface waters where they may give rise to impacts on the chemical water quality and ecological status of receiving waters as already recognised by the EU-Water Framework Directive. Leaking sewer pipes, land application of treatment residues and increased focus on soil-infiltration of storm water and wastewater further put the urban and peri-urban soil and groundwater resources at a potential risk that has only rudimentarily been assessed.

One associated topic involves methods for removal of xenobiotics in drinking water, waste- and storm water. Both "process innovation" (i.e. integration of chemical and biological processes and use of sequential reaction environments for enhancing biotransformation) and "technological innovation" (i.e. high rate bio film reactors and membrane reactors) are necessary to include. Fate of xenobiotics during the treatment and different approaches for modelling of the treatment processes also must be considered. Technological challenges are in focus and both low- and high-tech methods have to be covered, as well as centralised and decentralised strategies.

2. Treatment Techniques

In order to achieve the required ecological and chemical quality of surface water in 2015, additional discharge treatment steps will have to be implemented at WWTPs. This may require a different approach for each WWTP, depending on the occurrence of substances in the effluent, the configuration of the WWTP, the nature of the receiving water body and the contribution of the effluent load of contaminants to the total loading of the surface water. Treatment techniques that can potentially be applied to WWTPs to achieve the new effluent standards are membrane bioreactors, oxidative techniques, chemical precipitation techniques and adsorption techniques.

2.1. MEMBRANE BIOREACTOR

The MBR is an activated sludge system in which the sludge/water separation step takes place via micro- or ultra filtration membranes, instead of in secondary clarifiers. From a biological viewpoint the MBR is comparable to an optimized conventional activated sludge system with a post-treatment step by MF- or UF-membrane filtration; the difference in effluent quality is determined by the separation efficiency of the membrane system in comparison to a secondary clarifier. Biological techniques are effective for the removal of nutrients in the concentration range of a few mg/l. They are however not effective for the removal of organic contaminants to the level of μg/l concentrations, because these concentrations are too low to enable the growth of a biomass population, due to competition with the higher numbers of general heterotrophic bacteria.

2.2. OXIDATIVE TECHNIQUES

Oxidative techniques are used to "crack" organic compounds with the aid of strong oxidants such as ozone or hydrogen peroxide. They are applied for the oxidation of organic contaminants. Some oxidants can also be applied for disinfection. A specific branch of oxidation processes is the Advanced Oxidation Processes (AOP). By combining oxidation techniques, free radicals can be formed which cause the oxidation processes to proceed by a factor of 10 to several thousand times faster. Applicable combinations are ozone/hydrogen peroxide, ozone/UV and UV/hydrogen peroxide. An important advantage of chemical oxidation is that it is effective for contaminants in extremely low concentrations (μg/l). Oxidative techniques are potentially applicable for the removal of organic micro-contaminants from WWTP-effluent in the required concentration range. The relatively high concentrations of organic compounds in WWTP effluent require an effective pre-treatment. Due to the high costs of advanced oxidation in combination with UV, alternative combinations without UV are recommended to improve the feasibility of advanced oxidation for effluent treatment. The formation of by-products and their possible toxicity is a point requiring particular attention during application of oxidation.

2.3. CHEMICAL PRECIPITATION TECHNIQUES

Chemical precipitation is the addition of chemicals to water that bind with other substances in the form of a precipitate, which is subsequently removed. Examples of techniques that form a so-called "precipitate" are precipitation, coagulation and flocculation. Coagulation/flocculation is a suitable technique for the removal of suspended solids and colloidal material, to which micro-contaminants and heavy metals are often attached. The application of coagulation/flocculation in combination with filtration techniques can result in very effective removal of suspended solids. The technique is not suited to the removal of specific dissolved substances in the μg/l range. Disadvantages are the chemical consumption and subsequent chemical sludge production.

2.4. ADSORPTION TECHNIQUES

Adsorption refers to the attachment of substances in the water phase to a fixed surface. For the treatment of effluent is applicable activated carbon- the bonding of non-polar organic compounds due to Van der Waals forces; ion exchange: bonding of ions to specific charged groups on the surface of a synthetic resin; other adsorptive surfaces with some kind of affinity for organic and/or inorganic components.

Important criteria for the adsorbent material are a high specificity for the contaminant, such that low concentrations are attained and other components are not adsorbed, the surface area is as large as possible such that the maximum amount of adsorption occurs per unit adsorptive material, preferably capable of releasing the adsorbed components under controlled conditions via a technically and economically acceptable discharge.

3. Generation and Screening of Treatment Trains

Models for generation of treatment trains are aimed at generating different treatment alternatives, which involves database search, combinatorial search and inferential reasoning. These models are significant in the preliminary design stage as a number of treatment alternatives are to be generated and evaluated before selecting the best or near optimal treatment alternatives. For such an optimal solution is dependent on the generation of excellent and different alternatives.

In recent years with emphasis on sustainability and protection of the environment a number of criteria have been addressed. Most widely used criteria are cost of treatment, effluent quality achieved, reliability, land required, easy of operation and maintenance, resource requirement, quantity and quality of sludge produced, impact on the environment and adaptability.

Decision support systems have been defined as interactive computer based systems, which facilitate complex decision making by providing reasoning support through database and mathematical optimization based on the strength of human designers. Decision support system is a computer program that assists individuals or groups of individuals in their decision processes, supports rather than replaces judgment of these individuals and improve the effectiveness rather than efficiency of decision process. Advantages of using a decision support system for generation, evaluation and optimization of treatment trains are that it can be used for analysis of “what if” scenarios, the user can change assumed design parameters or removal efficiencies and see the difference and costs almost immediately, it has a knowledge base that the user can access and update easily through a user interface, an increasing number of alternatives can be examined using a decision support system, it has a structured approach to generation, evaluation and optimization and results will be consistent. It can be used for ad-hoc analysis of treatment trains, it can be less time consuming than a traditional "rule of thumb" approach and it can be used to optimize treatment train selection.

4. Model Development

A process simulation model XENA has been developed which was used in combination with an integrated optimisation engine to allow a range of design possibilities for removal of Xenobiotics to be explored. The model includes a computational module for wastewater treatment trains and a knowledge base. The computational modules are used to calculate the performance of user defined system alternatives, utilizing the information contained in the knowledge base that includes rules for generation of treatment trains, design, cost and evaluation criteria information. XENA is based on a hydraulic/process simulation model WTRNet (Joksimovic 2006). The development of WTRNet was aimed at providing a DSS that will overcome some of the limitations that appear in currently available decision support tools, by including the following features:

- ✓ The ability to include multiple treatment facilities within a wider water reuse scheme in the evaluation, where the treated wastewater could be upgraded at several sites to meet the specific quality requirements of multiple users (distributed treatment).
- ✓ Provide a completely open modelling environment, that will allow users the flexibility in terms of editing the information contained in the model knowledge base, and adding information to the knowledge base (e.g. unit processes and their characteristics, pollutants to be considered, use types and quality requirements, rules for combining unit processes in a treatment train, etc.),
- ✓ Provide suggestions for complete treatment trains based on the influent quality (or current level of treatment provided in the case of existing wastewater treatment facilities) and quality requirements for "standard" end uses of reclaimed water,

The simulation model allows planners of WWTP's to explore a large number of design alternatives in an efficient manner, for the purpose of identifying the most promising schemes.

5. Model Description

XENA is a simulation and optimization software which is used to evaluation of treatment trains.

5.1. SOFTWARE STRUCTURE

The structure of the software includes three main components. These components are: Control module, which is used as a coordinator between a knowledge based model and computational modules, the treatment performance module, and the Control Module is used along with a graphical user interface (GUI) for control of input data and display of results.

5.1.1. *Knowledge Base Model*

The Knowledge Base contains the following data: Unit Processes detailed information – suggested pre-cursors or post-cursors of unit process, Pollutant Removals for the basic pollutants, Costing Data (Capital Cost, EM Cost, O&M Cost), Resource Data (Land requirements, Labour requirements, Sludge Production, Concentrate Production and Energy Consumption) and Evaluation Criteria Scores.

5.1.2. *Treatment Performance Module*

The treatment performance has been developed with functionality to perform the evaluation of user-selected combination of unit processes in a treatment train. The evaluation of alternative treatment trains includes sizing the unit process, and calculating the effluent quality achieved by the current treatment train, pollutant percent removed, evaluation criteria scores, costs and resources.

6. Optimization

Two optimization techniques are used in the software of XENA. The first one is exhaustive enumeration and the second one is Genetic Algorithms (GA) optimization. Using exhaustive enumeration, all possible design alternatives are explored, with respect to alternative treatment trains that satisfy their requirements. Genetic algorithms are a set of guided search procedures based on Darwin's theory of natural selection. The basic idea of GA is maintain a set of solutions, which evolve over time through process of survival of the fittest similar to the population genetics in nature. The first step involved in running GA is the random generation of a population. A random number is generated by a built in random generator in the programming language. The next step is evaluation which involved calculation of a fitness score for each solution. The fitness function is a means of rating how good or bad an individual system is, as compared to others in the population. Better individuals have higher chances of survival and reproduction.

7. Conclusions

The process simulation model presented here provides a comprehensive framework for analysis of treatment trains for removal of micropollutants. Further work will concentrate on knowledge based completion.

Acknowledgements

The authors would like to express their gratitude to the Czech Ministry of Education for funding this work within the COST project on "Methods for Xenobiotic Treatment in the Urban Water."

References

Joksimovic, D., Kubik, J., Hlavinek, P., Savic, D. and Walters, G.: Development of an Integrated Simulation Model for Treatment and Distribution of Reclaimed Water (2006), pp. 9-20

Dinesh, N.: Development of a Decision Support System for Optimum Selection of Technologies for Wastewater Reclamatio n and Reuse (2002), University of Adelaide, Adelaide, Australia, pp. 479

Joksimovic, D., Kubik, J., Hlavinek, P., Savic, D. and Walters G.: Development of a Simulation Model for Water Reuse Systems (2004), (published on CD), Marocco, September 2004

Loughlin, D.H., Doby, T.A., Ducoste, J.J. and de los Reyes, F.L.: System-Wide Optimization of Wastewater Treatment Plants Using Genetic Algorithms (2001), ASCE, Orlando, Florida

Hlavinek, P., Kubik, J., Case Study In The Czech Republic - Optimization Of Treatment Train For Water Reuse Schemes (2007), proceedings NATO Advancer Research Workshop "Wastewater Reuse - Risk Assessment, Decision-Making and Environmental Security", Istanbul, Turkey, ISBN: 978-1-4020-6026-7

STOWA: Exploratory Study for wastewater Treatment Techniques and European Water Framework Directive, 34/2005

SIMULATION OF THE RADON FLUX ATTENUATION IN URANIUM TAILINGS PILES

MARIA DE LURDES DINIS, ANTÓNIO FIÚZA
Geo-Environment and Resources Research Center (CIGAR)
University of Porto, Engineering Faculty, R. Dr. Roberto Frias, 4200-465, Porto, Portugal. mldinis@fe.up.pt; afiuza@fe.up.pt

Abstract. Tailings wastes are generated during the milling of certain ores to extract uranium and thorium. In the recent past uranium mill tailings consisted of fine-grained sand and silt materials, usually disposed in large piles in an open air area. Radium is probably the most hazardous constituent of uranium tailings. It produces radon, a radioactive gas which can easily spread into the environment. Airborne radon decays into a series of short half-life products that are hazardous if inhaled. Tailings also emit gamma radiation which can increase the incidence of cancer and genetic risks. Post closure and site rehabilitation involves, among other situations, controlling and estimating radon release from the surface of the tailings pile. Generally the primary cleanup method consists of enclosing the tailings with compacted clay or native soil to prevent the release of radon and then covering this layer with rocks and vegetation. This implies a cover design and placement which will give long term stability and control to acceptable levels of radon emission and gamma radiation, preventing also erosion and water infiltration into the tailings.

An algorithm based on the theoretical approach of diffusion was developed to estimate radon attenuation originated by a cover system placed over the tailings pile and subsequently the resulting concentration in the breathing atmosphere. The one dimensional steady-state radon diffusion equation was applied to a porous and multiphase system to estimate the radon flux from the tailings to the surface. The thickness of a cover that limits the radon flux to a stipulated value was performed for a particular contaminated site. The efficiency of the cover attenuation was evaluated from the comparison with the resulting radon concentration in the absence of any cover system.

Keywords: radon, cover system, tailings, radium, flux, diffusion

M.D. Annable et al. (eds.), Methods and Techniques for Cleaning-up Contaminated Sites, 129–136.

1. Introduction

Because uranium mills use chemical processes to selectively remove uranium from the ore, contaminants in uranium mill tailings include in most cases radium and thorium, which are the dominant radioactive materials in mill tailings. Tailings also contain small residual amounts of uranium that were not extracted during the milling process. Uranium mill tailings were normally dumped as sludge in special ponds or piles, where they were abandoned. Most of the radioactive contaminants are still present in these materials which have been brought by mechanical and chemical processes to a condition where the contaminants are now much more mobile and thus susceptible to migrate into the environment.

Radon-222 gas emanates from the tailings being continuously produced from the radioactive decay of radium-226. Some of this radon escapes from the interior of the pile and may be quickly spread with the wind being dispersed in the environment. To avoid environmental contamination, a rehabilitation plan should initially promote the confinement of the radioactive wastes, followed by a geotechnical stabilization of the pile and then by a multilayer covering system placed over the surface of the tailings disposal.

This work is based in a mathematical model for simulating the radon flux attenuation by designing a multilayer cover system and monitoring its effect by predicting the respective concentration at a defined mixing height. The model estimates the radon flux released from the radium content in the tailings and the radon concentration at the breathing or mixing height. The efficiency of the cover system designed is evaluated by comparing the resulting radon concentration at the breathing height considering an inexistent cover system.

2. Methods and Materials

A methodology is proposed to describe the radon exhalation attenuation from a uranium tailings disposal and it may be divided into two main sub-models:

1. The estimation of the flux released from the tailings to the air breathing zone, considered with or without a cover system, estimated by the diffusion equation applied to a multiphase media.
2. The estimation of the corresponding radon concentration predicted by a box model formulation.

Radon emanation has been addressed mostly as diffusive transport from the soil. The main soil properties and cover materials control the radon migration from the subsoil to the surface. In order to optimize a potential efficient cover design, soil and air transport processes in the tailings and in the cover layers were investigated.

2.1. RADON EXHALATION

The movement of radon in the soil or in the tailings can be described by the diffusion coefficient, D ($cm^2.s^{-1}$), which can be estimated by an empirical correlation with the fraction of saturation, m (Rogers and Nielson, 1984):

$$D = 0{,}07\, e^{\left[-4\left(m - m\,\varepsilon^2 + m^5\right)\right]} \quad (1)$$

The radon migration from the subsoil to the surface is controlled by the tailings and cover materials properties such as porosity (ε), moisture (θ), and the degree of saturation (m). This last parameter depends on the commonly measured moisture and may be estimated with the dry bulk density, ρ ($g.cm^{-3}$), and the specific gravity, g ($g.cm^{-3}$) of the diffusing media (tailings or cover) (Rogers and Nielson, 1984):

$$m = 10^{-2} \cdot \frac{\theta}{\frac{1}{\rho} - \frac{1}{g}} \quad (2)$$

The release mechanism from the soil is based in the principles of radon diffusion across a porous medium which may be estimated with the one-dimensional steady-state radon diffusion equation:

$$D \frac{\partial^2 C}{\partial x^2} - \lambda C + \frac{R \rho \lambda E}{\varepsilon} = 0 \quad (3)$$

The first term of this equation defines the diffusion transport given by the Fick's law in a one-dimensional form, which means that the diffusing substances are proportional to the concentration gradient ($J = -D.\partial C/\partial x$). The second term represents a first order decay kinetics ($dJ/dx = -\lambda.C$) and the last term represents the radon generation from the radium decay ($R.\rho.\lambda.E/\varepsilon$).

The necessary parameters for solving this equation are the radon diffusion coefficient, D ($m^2.s^{-1}$), the radon decay constant, λ (s^{-1}), the radium concentration in the pores space, C ($Bq.m^{-3}$), the radium concentration in the tailings, R ($Bq.kg^{-1}$), the bulk density of the dry material, ρ ($kg.m^{-3}$), the radon emanation coefficient, E (dimensionless), the total porosity, (ε) (dimensionless) and the moisture, (θ) (dimensionless).

The generic solution of the diffusion equation (3) gives the radon flux released, J ($Bq.m^{-2}.s^{-1}$), which may be applied in 3 main situations (figure 1): (i) the flux to the atmosphere without cover trapping, J_t; (ii) the flux through a simple cover system, $J_{C(1)}$ and (iii) the flux through a multilayer cover system, $J_{C(2)}$.

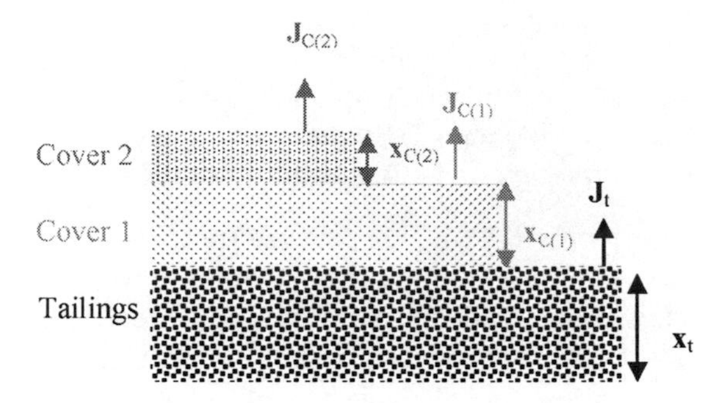

Figure 1. Parameters for a multilayer cover system, (NUREG, 1984).

The solution of the generic diffusion equation for a homogeneous medium represents the flux release, J_t ($Bq.m^{-2}.s^{-1}$), from the tailings with a thickness of x_t (m) without a cover system and is given by (Rogers and Nielson, 1984):

$$J_t = R\rho E\sqrt{\lambda D_t}\tanh\left(\sqrt{\frac{\lambda}{D_t}}x_t\right) \tag{4}$$

For a two media problem, tailings (t) and a homogeneous cover material (c), the solution for the generic equation, J_c, is given by the equation (5) with $b_i = \sqrt{\lambda / D_i}$ (m) (i = c or t) and $a_i = \varepsilon_i^{\,2} D_i[1-0{,}74\,m_i]^2$ ($m^2.s^{-1}$) (Rogers and Nielson, 1984):

$$J_c(x_c) = \frac{2\cdot J_t e^{-b_c x_c}}{\left[1+\sqrt{a_t/a_c}\,\tanh(b_t x_t)\right]+\left[1-\sqrt{a_t/a_c}\,\tanh(b_t x_t)\right]e^{-2\cdot b_c x_c}} \tag{5}$$

And the cover thickness, x_c, for a stipulated flux is obtained by rearranging the equation (5) for the tailings and the cover parameters (Rogers and Nielson, 1984):

$$x_c = \sqrt{\frac{D_c}{\lambda}}\ln\left[\frac{2\cdot J_t / J_c}{\left(1+\sqrt{a_t/a_c}\,\tanh(b_t x_t)\right)+\left(1-\sqrt{a_t/a_c}\,\tanh(b_t x_t)\right)\left(Jc/Jt\right)^2}\right] \tag{6}$$

2.2. RADON CONCENTRATION

For the estimation of the radon release to the breathing zone and the corresponding concentrations, a box model is used (figure 2). In a box model formulation the contamination source is defined by an emission area generating a constant emission rate (ϕ).

The box volume is defined by its length (L), width (w) and the mixing height (h) and inside, the concentration (C) is spatially homogeneous and constant in time resulting from the assumption of a complete mixing inside the compartmental box. Radon is directly diluted into the local air existing in the breathing zone above the contaminated source, and is carried away by the atmospheric circulation across it. The value for the wind speed (u) used in the calculation of radon dilution is matched by the average annual values through the mixing zone.

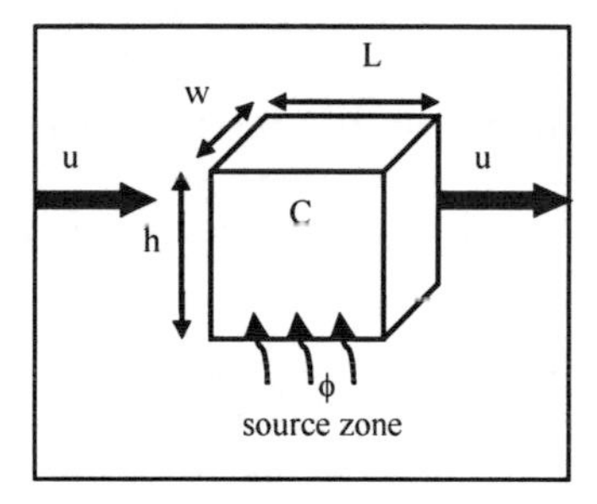

Figure 2. Box model for the radon concentration estimative at the breathing height, h.

In this type of model formulation, a mass balance concept is implicit:

$$V\frac{dC}{dt} - \phi A - uSC \tag{7}$$

and as consequence of a steady state assumption, we have that the pollutant concentration (C) is constant in time:

$$V\frac{dC}{dt} = 0 \tag{8}$$

that the mass flow rate entering (ϕA) into the box is equal to the flow rate leaving the box (uSC):

$$\phi A = uSC \tag{9}$$

and that the residence time (τ) is defined as the average amount of time that a particle of material remains in the box:

$$\tau = \frac{CV}{\phi A} \tag{10}$$

3. Case Study

3.1. PROBLEM DESCRIPTION

The intensive and extensive uranium mining and processing operations in Portugal has left a legacy of considerable environmental contamination. The uranium exploitation in Portugal began in 1913 and ended in 2000. During this period the uranium was mined in 62 different places. One particular place is Urgeiriça where the extensive exploitation and treatment of the uranium ore in this mine and others from the same region, has led to an accumulation of large amounts of solid wastes: about 4 million tons of rock material was routed into natural depressions confined by dams that cover an area of about 13.3 hectares. The radon exhalation from the tailings is potentially one of the main sources of contamination for nearby areas.

The Urgeiriça tailings pile was considered as a case study. The area and the average thickness of the mil tailings deposit were estimated at 12 hectares and 14 m, respectively. The radium concentration in the mill tailings was measured and the resulted value for ^{226}Ra was 12,900 Bq/kg. A value of 0.24 was assumed for radon emanation coefficient. A total porosity of 0.37 was considered for the tailings materials and a value of 1,67 g.cm^{-3} was used for bulk density (Pereira et al., 2004).

The radon flux in the surface of the pile depends on the diffusion coefficient, which is estimated with the moisture content measured in the tailings, the total porosity and the ^{226}Ra activity in the tailings. An average value of $9{,}2 \times 10^{-7}$ m^2.s^{-1} was obtained for radon diffusion coefficient in the tailings (Pereira et al., 2004).

A cover design of 5.15 m average thickness composed by sand, clay and gravel was proposed on the basis of the plan rehabilitation for the site (Pereira et al., 2004). An average value of 2.2 g.cm^{-3} was considered for the cover bulk density. A total porosity of 0.30 was considered for the cover material and the diffusion coefficient was estimated at 5×10^{-7} m^2.s^{-1}.

3.2. RESULTS

The methodology used has as outputs: i) the radon flux without cover and the resulting concentration at one m above the ground considered the breathing height; ii) the radon flux attenuation produced by the cover system proposed and iii) the estimation of a cover thickness that allows a radon flux less than to a value stipulated as safe or permissible.

The value obtained for the average radon flux without cover was $J_t = 7.19$ Bq.m^{-2}.s^{-1} and the resulting concentration at 1 m above the ground at the site was $C_0 = 547.4$ Bq.m^{-3}. The average value measured in the Urgeiriça tailings pile for the radon concentration was 557 Bq.m^{-3} (Exmin, 2003). This corresponds to a dose of 5.2 mSv/year, resulting from the exposure to radon in outdoor air, assuming that the receptor spends 1760 h outdoor in a year (a 20% outdoor residence time), with an equilibrium factor between ^{222}Rn and its decay products of 0.6 and for a dose conversion factor of 9 nSv.h^{-1} per Bq.m^{-3} (Grasty and LaMarre, 2004).

The radon flux attenuation originated by the cover system proposed with a thickness of 5,15 m, a porosity of 0,30 and a diffusion coefficient of $5{,}0 \times 10^{-7}$ m^{2}.s^{-1} is equal to $J_c = 187$ μBq.m^{-2}.s^{-1} and the corresponding concentration at the breathing height is equal to $C_0 = 0{,}0142$ Bq.m^{-3}, which is negligible.

We tested the same materials efficiency considering that the tailings pile will be covered with 0,5 m of clay plus a layer of overburden to achieve a surface flux less than the permissible one, which was considered to be 0.74 Bq.m^{-2}.s^{-1} or 20 pCi.m^{-2}.s^{-1} (EPA, 1983). We considered the clay cover layer and the same diffusion coefficient used in the previous example. For the overburden layer we considered a diffusion coefficient of 2.2×10^{-6} m^{2}.s^{-1} and a porosity equal to 0.37.

The radon flux attenuation through the clay component cover is $J_{c1} = 2.63$ Bq.m^{-2}.s^{-1} with a concentration of 200,23 Bq.m^{-3}. The diffusion coefficient for the new source term (tailings plus clay layer) is estimated at 6.98×10^{-7} m^{2}.s^{-1} and this value is used to estimate the second layer cover thickness which only allows the exhalation of the stipulated flux. The value obtained for the second layer is $x_{c2} = 1.52$ m which gives a total cover thickness of 2.02 m. The resulting concentration at the breathing height is 56.34 Bq.m^{-3} which corresponds to a dose, in the same previous conditions, of 0.53 mSv/year. This dose is less than 1 mSv/year, the limit derived from the European guidelines concerning the exposure of the general public to artificial radionuclides.

4. Conclusions

The placement of an engineered cover designed to isolate the tailings and any other material is efficient in attenuating radon emanation to a safe level. It also acts as a barrier preventing rainwater infiltration into the tailings pile as well as wind and water erosion. The optimized cover construction proposed to be built reduces the radon exhalation rate, and subsequently the radon concentration at the site, to negligible values once the natural background is less than 30 $Bq.m^{-3}$. If the desired performance criterion is to achieve the radon standard for ^{222}Rn emission rate from the surface of inactive uranium mill tailings piles (EPA, 1983), a lower thickness cover could be used. In the example given, the two layers of cover produces reduction of the radon flux and concentration at the breathing height by a factor of nearly 10.

When the rehabilitation plan is implemented, in particular the cover system proposed, monitoring and maintenance actions should be carried out to achieve long term efficiency.

Acknowledgement

This work has been carried out with the financial support of Foundation for Science and Technology (FCT - MCES).

References

EPA, Environmental Protection Agency (1983), "Final Environmental Impact for Standards for the Control of Byproduct Materials from Uranium Ore Processing", 40 CFR 192, Report No. EPA/520/1-83-008-1.

Exmin, Companhia de Indústria e Serviços Mineiros e Ambientais, SA. (2003), "Estudo Director de Áreas de Minérios Radioactivos – 2.ª fase".

Grasty, R. L. and LaMarre, J. R., "The Annual Effective Dose From Natural Sources of Ionising Radiation in Canada", Radiation Protection Dosimetry (2004), Vol. 108, No. 3, pp. 215-226 DOI: 10.1093/rpd/nch022.

Pereira, A. J. S. C., Dias, J. M. M., Neves, L. J. P. F. & Nero, J. M. G. (2004), "Modelling of the long term efficiency of a rehabilitation plan for a uranium mill tailing deposit (Urgeiriça – Central Portugal)", XI International Congress of the International Radiation Protection Association.

Rogers, V. C. and Nielson, K. (1984), "Radon Attenuation Handbook for Uranium Mill Tailings Cover Design", U.S. Nuclear Regulatory Commission, NUREG/CR-3533.

BIOCHEMICAL OXIDATION – A PATHWAY FOR AMMONIA REMOVAL FROM AQUATIC SYSTEMS

MARIA SANDU*, P. SPATARU
Institute of Ecology and Geography, Chisinau, 5 , str. Tudor
TATIANA ARAPU, T. LUPASCU
Institute of Chemistry, Chisinau, 3, str. AcademieiMD-2028

Abstract. The river network in the Republic of Moldova is the final receptor of most runoff and wastewater with insufficient treatment level. About 1/3 of pollutants, including ammonia compounds, still persist after treatment, is discharged into natural waters. In the preceding thousand years the course of all small rivers was straighten and basins where drainage. The self-purification process and oxidation of ammonia to nitrate by microorganisms has in their waters a low rate. This paper presents the results of researches that include laboratory modelling of the role of diverse substrates in biochemical oxidation of ammonia as a Pathway for Ammonia Removal from Aquatic Systems. The biomineralization process of ammonium ions was simulated in small rivers water in the presence of diverse substrates (polyethylene film, sandy soil, stone, gravel, active carbon). Biochemical transforming of nitrite produced by ammonia oxidation is reduced to nitric oxide, nitrous oxide and possibly to nitrogen gas. Particularly the gravel accelerates the rate of process more than 2-3 times.

Keywords: water; nitrification; oxidation; ammonia; nitrite

1. Introduction

1.1. STATE OF AQUATIC RESOURCES

As in any other country, the river network in the Republic of Moldova is the final receptor of most runoff and wastewater either household or industrial; the treatment level of wastewaters is insufficient and 1/3 of pollutants, including

* Maria Sandu, Institute of Ecology and Geography, Chisinau, 5, str. Tudor

M.D. Annable et al. (eds.), Methods and Techniques for Cleaning-up Contaminated Sites, 137–143.

ammonia compounds, pesticides, some heavy metals still persist after treatment, which is discharged into natural waters. The concentrations of pollutants, such as ammonia ions, regularly exceed the MACs (Lozan, 2003). The content of analyzed metals in the Nistru river: *Cu -1.9-18.9* $\mu g/dm^3$, *Zn - 18.8-178* $\mu g/dm^3$, *Cd - 0.9-4.1*$\mu g/dm^3$, *Cr - 0.3-1.7* $\mu g/dm^3$. The highest concentrations are recorded downstream of the confluence with river Bac (carrying the waste water from the capital city Chisinau) and upstream the cities of Soroca, Tighina and Tiraspol: zinc - 122 - 176 $\mu g/dm^3$, copper 26,8 $\mu g/dm^3$ – 32 $\mu g/ dm^3$ and so on. The poor water quality in small internal rivers is an issue of permanent concern. The concentrations of ammonium, nitrites, phenols and copper regularly exceed the MACs. Water pollution index (ratio of the substance discharged content and the discharge limit value) for Bac river is given in Figure 1. Ammonium concentrations of 17.0 mg/dm^3 as N (43.6 MAC) up to 46.5 mg/dm^3 as N (119 MAC) were found downstream Chisinau, maintaining their high values up to the confluence with the Nistru river.

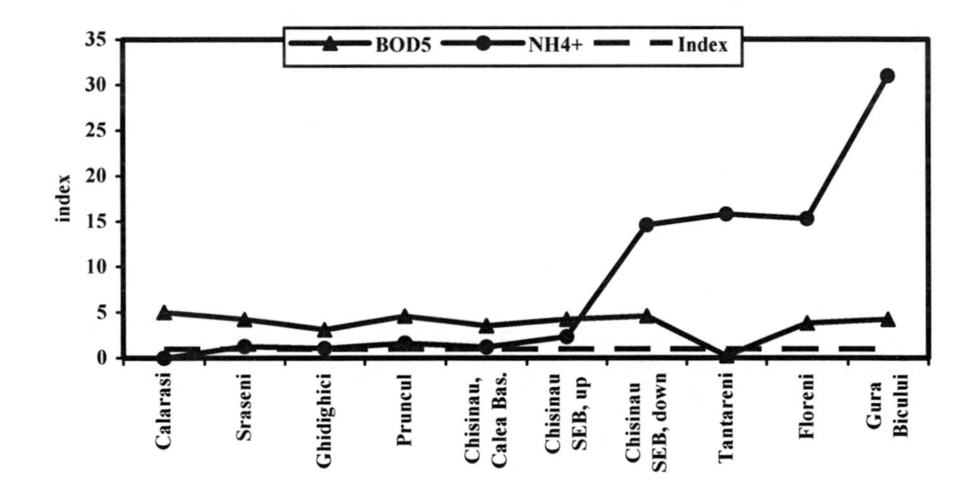

Figure 1. Spatial evolution of the Bic river water quality

Non-polluted natural waters contain 0,02-0,05 mg/dm^3 of ammonia. The discharge of ammonium into the surface waters is higher for river Prut. All small rivers have low flows and a poor natural self-purification and nitrification capacity (Lozan, 2004).

Ammonium ions can be formed during the anaerobic reduction of nitrates and nitrites. Its presence in unpolluted waters is connected with biochemical decomposition of proteins. An increase in ammonium is observed when aquatic organisms are dying off.

1.2. METHODS FOR THE REMOVAL OF AMMONIA

A technology for the removal of ammonia is using selective catalytic oxidation to nitrogen and water. This process which provides an efficient, stable, simple and selective purification of large gas emissions can be applied both in low and high concentrations of ammonia removal. It is also possible to selectively oxidize ammonia to nitrogen in the liquid phase. The catalytic oxidation of ammonia is one of important heterogeneous catalytic processes (Gang 2000, Gang 2002).

Another way of NH_3 oxidation is a two-step biochemical process catalyzed by ammonia monooxygenase (AMO) and hydroxylamine oxidoreductase (HAO). AMO catalyzes the oxidation of NH_3 to NH_2OH ($NH_4^+ + 1/2\ O_2 \rightarrow NH_2OH + H^+$) and HAO catalyzes the oxidation of NH_2OH to NO_2^- ($NH_2OH + O_2 \rightarrow NO_2^- + HOH + H^+$) (Arp, 2002). The microorganisms involved are called the ammonia oxidizers. There are four recognized genera (Nitrosomonas, Nitrosospira, Nitrosococcus, Nitrosolobus). Nitrosomonas is the most extensively studied and usually the most numerous in soils. Nitrosospira is an aquatic nitrifier. In the result of biochemical Oxidation 66 kcal of energy are liberated per gram atom of ammonia oxidized.

Second step is biochemical nitrite oxidation ($NO_2^- + 1/2\ O_2 \rightarrow NO_3^-$) by means of Nitrobacter. As a result 18 kcal of energy is liberated per gram atom of nitrite oxidized.

Nitrite-oxidizing bacteria are found in aerobic, but occasionally also in anaerobic, environments where organic matter is mineralized.

Ammonia-Oxidizing Bacteria are aerobic but can grow at reduced oxygen partial pressure. The organisms are chemoautotrophs, growing with ammonia as the energy substrate and CO_2 as the main carbon source. Generally, the optimum growth temperature is 30°C and the optimum pH is between 7.5 and 8.0. Species are distributed in a great variety of soils, oceans, brackish environments, rivers, lakes, and sewage disposal systems. Nitrifiers need oxygen to perform their task (aerobic respiration). Nitrate is the final product after completion of the biochemical oxidation.

Nitrification, the oxidation of ammonia to nitrate by microorganisms, is a key process in the global nitrogen cycle, resulting in nitrogen loss from ecosystems, eutrophication of surface and ground waters, and the production of atmospherically active trace gases. Nitrifying bacteria work either at full capacity or drift into a resting phase.

Some substances will kill bacteria, such as phenol, surface-active compounds and others.

The process occurs most rapidly between 20 and 30 degrees. It is influences by content of water pollutants. It was established the influence of substances (humic acids, lindan, dieldrin, heavy metals, urea, detergents, phenol, petroleum products) on biochemical oxidation of ammonia. The quantity equal to 1 and 2 maximal admissible concentration (MAC) of anionic, unionic and its mixture aid development of heterotrophic bacteria, unlike cationic ones and there mixtures with anionic and unionic surface-active substances. Unionic surface-active compound consumes more intensive dissolved oxygen. The content of 5 MAC cationic substances and rather there mixtures provoke intensive development of bacteria (Sandu, 1993), (Sandu, 2006), (Spataru, 2003).

It has been suggested that in bacteria nitrification is used as a sink for excess reducing power generated by oxidative metabolism. It has been shown that many heterotrophs couple ammonia oxidation to aerobic denitrification. In this process the nitrite produced by ammonia oxidation is reduced to nitric oxide, nitrous oxide, and possibly to nitrogen gas. Ammonia oxidizing bacteria extract energy for growth from the oxidation of ammonia to nitrite. Other inorganic nitrogen compounds, including NO, N_2O, NO_2, and N_2 can be consumed and/or produced by ammonia-oxidizing bacteria. Nonetheless, many questions remain regarding the role of these pathways and pollutants in ammonia oxidizers (Arp, 2002).

So, this paper presents the results of researches that include laboratory modelling (glass bottles) of the nitrification capacity of small rivers water and the role of diverse substrates in biochemical oxidation of ammonia.

2. Results and Discussion

2.1. NITRIFICATION CAPACITY OF SMALL RIVERS WATER

Specifically for small rivers is natural variation of water quality and temporary variation of the debit. The evolution has unregulated concentrations comportment without correlation to river debit. In such case is talking about external chance contribution or variable biological phenomena in water that disconnected with the debit and other factors.

Nitrification capacity of small rivers water was investigate by means of laboratory modelling (glass bottles) with employment of natural waters (Sinelnicov, 1980), (Soares, 1988). Initial concentration of ammonia ions was creating by adding 2 mg/dm^3 of NH_4^+. In view of the fact that the change of microorganisms content in function of the speed has not a linear character analytical settlement of problem was realized by control of NH_4^+, NO_2^-, NO_3^-, O_2 and pH (Leitte, 1971), (UMWQI, 1987).

Out by experimental obtained results it was established that $NH_4^+ \rightarrow NO_2^-$ process in modeling of river Raut (Baraboi), Cubolta (Garbova), Ciuhur (Ocnita) and lace in Hlinoaia water it takes place in the same way as river Nistru water and lasts 6-10 days (Figure 2), whereas in river Racovat water only 40-77% of ammonium ions are oxidizes. Even after 15-20 days in solutions remained 7-20% of NH_4^+.

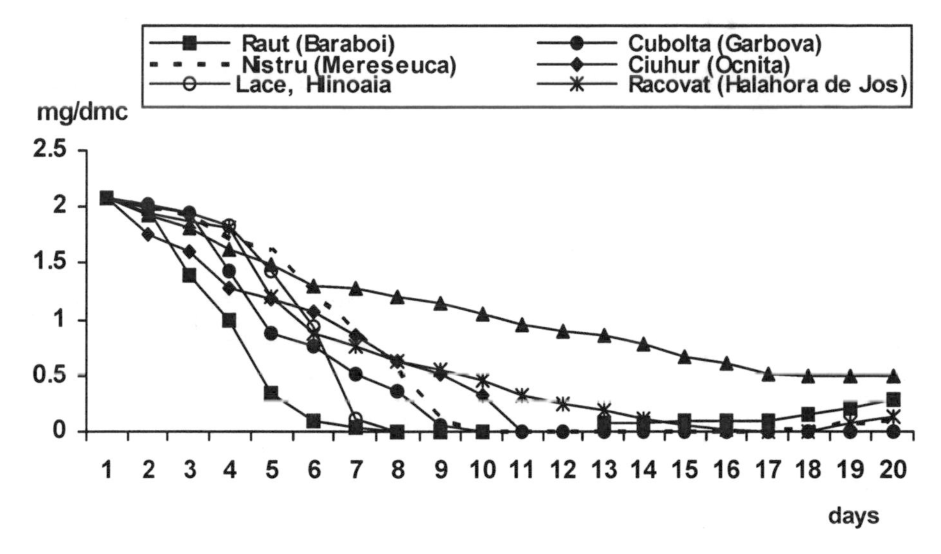

Figure 2. Dynamics of process $NH_4^+ \rightarrow NO_2^-$ in small rivers water

The process $NO_2^- \rightarrow NO_3^-$ continues 10 days in river Raut and lace from Hlinoaia water; in Cubolta river – 13-14 days and Nistru – 16-17 days. Nitrites don't disappear in Racovat and Ciuhur rivers water for 20 and more days.

So, one of natural waters problems in the Republic of Moldova is toxic manifestations of the pollutants on biochemical oxidation of ammonia ions that may vary depending not only on the dose but also on their nature.

2.2. THE ESTIMATION OF THE ROLE OF SUBSTRATES IN NITRIFICATION

Nitrification process in water was investigated by means of laboratory modelling (glass bottles) (Sinelnicov, 1980), (Soares, 1988). Water samples for modeling were collected from rivers Nistru. Initial concentration of ammonia ions was creating by adding 2 mg/dm^3 of NH_4^+.

Proceed from of obtained results of researches it was estimated the role of diverse substrates (polyethylene film, sandy soil, stone, gravel, active carbon) in biochemical transforming $NH_4^+ \rightarrow NO_2^- \rightarrow NO_3^-$.

The laboratory modeling demonstrate that various type of substrates and aeration stimulates biochemical oxidation $NH4^+ \rightarrow NO_3^-$. The decreasing in experiment of ammonia ions content in the presence of studied substrates denote that gravel is the most active and accelerate the rate of process more than 2-3 times in comparison with the control - Water, Nistru (N)–daylight-(L) (Figure 3).

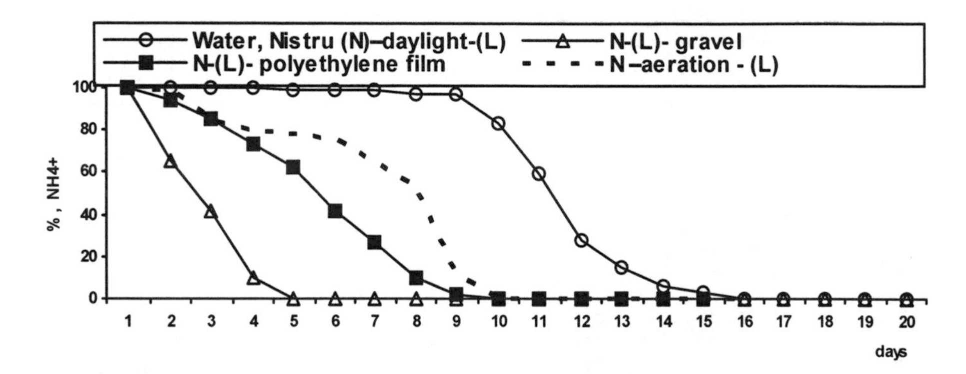

Figure 3. Evolution of NH_4^+- content in model solutions

The second stage of nitrification ($NO_2^- \rightarrow NO_3^-$) in absence of gravel lasts about 18–20 days; in its presence – 6 days. Aerial treatment speeds up the oxidation of ammonia ions in system till 9 days (Figure 4). It is to take note that gravel contributes to elimination of ammonia out of water under various forms, because at the end of the experiment the content of nitrates that was in initial water increased insignificant.

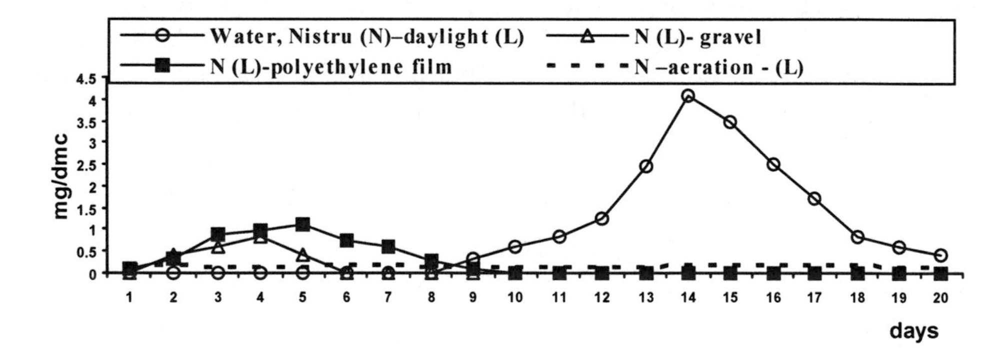

Figure 4. Evolution of NO_2^- content in the model

The value of pH in experiment increases a little (from 8,15 – 8,2 till 8,3 – 8,9).

Gravel substrate is proposed as measure for improving of water ecological state in small rivers and for wastewater biological treatment plants. That proposal is as argument for return to natural state of small rivers in the Republic of Moldova and for improving wastewater biological treatment plants activity (Scientific Report, 2005).

3. Conclusions

1. Small rivers in the Republic of Moldova have a poor nitrification and self-purification capacity. A problem for water is toxic manifestations of the pollutants on biochemical processes as oxidation of ammonia ions.
2. The gravel contributes to elimination of ammonia ions out of water and it is proposed as measure for improving of water state, self-purification in small rivers and for wastewater biological treatment plants.

Reference

Arp D.J., Sayavedra-Soto L.A. and Hommes N.G. (2002) Molecular biology of ammonia oxidation in *Nitrosomonas europaea*. Mini-Review. Archives of Microbiology (178), pp. 250-5.

Gang L., Anderson B.G., van Grondelle J. and van Santen R.A. (2000) NH3 oxidation to nitrogen and water at low temperatures using supported transition metal catalysts. Catalysis Today, 61, 179.

Gang L. (2002) Catalytic Oxidation of Ammonia to Nitrogen. China. Eindhoven : Technische Universiteit Eindhoven, Proefschrift.

Leitte A.D. (1971) Water quality criteria data book. Vol. 2. Inorganic chemical pollution of fresh water. 273 p.

Lozan R., Holban V., Ungureanu D., Usatai M. (2004) Water Pollution Sources, Republic of Moldova State of the Environment. Report, 2003. - Chisinău, pp. 31-32.

Sandu M., Rusu M., Lozan R. and Ropot V. (1993) The influence of some organic substances on biochemical oxidation of ammonia ions. Simp. "Omul şi mediul înconjurător". Iaşi, Romania. pp. 68-69.

Sandu M., Spataru P., Lupascu T., Arapu T. and Dragutan D. (2006) Evolution of nitrification process in natural waters. Conf. "Ecology and Environment protection – research, implementation, management". Chisinau, pp. 257-260.

Sinelnicov V. E. (1980) Mecanism of water body self-purification. Stroiizdat, 112 p.

Soares, M.I.M.; Belkin, S. and Abeliovich, A. (1988) Biological Groundwater Denitrification: Laboratory Studies, Water Science Technolgy Vol. 20, pp. 189-195.

Spataru P., Sandu M., Lupaşcu T., Dragalina G., Arapu T. and Lozan R. (2003) The influence of petroleum products on biochemical oxidation of ammonia ions, Bul. A.S.M. Seria st. biol. and chem., Nr. 2(291), pp. 128-130.

Unified methods of water quality investigations. p. I, vol. 1, M.. 1987, 907 p.

Scientific report. Evaluation of ecological state of water resources. Elaboration of methods for purification of liquid systems. National Institute of Ecology of the RM, Chisinau, 2005.

BIOBEDS - BIOTECHNOLOGY FOR ENVIRONMENTAL PROTECTION FROM PESTICIDE POLLUTION

MARÍA DEL PILAR CASTILLO,
LENNART TORSTENSSON
Swedish University of Agricultural Sciences, Department of Microbiology, Box 7025, SE-750 07 Uppsala, Sweden

Abstract. Point sources of pesticides, for instance frequently occurring at the filling of spraying equipment, are one of the most dominant reasons for pesticide pollution of surface and ground waters today. This contaminant risk can be minimized by using biobeds. Biobeds are facilities intended to retain and degrade pesticide spills. In its original design they consist of a biomixture, a clay layer at the bottom and a grass cover on the surface. The typical Swedish biomixture consists of straw, topsoil and peat (50-25-25 % v/v). The straw stimulates the growth of lignin-degrading fungi and the formation of ligninolytic enzymes (such as manganese and lignin peroxidases), which can degrade many different pesticides. The soil provides sorption capacity and other degrading microorganisms and the peat contributes to high sorption capacity and also helps to regulate the humidity of the system. A grass layer covering the biobed also helps to keep the correct humidity and can be used as an indicator revealing pesticide spills. The clay acts as an impermeable layer at the bottom. More than 1500 biobeds are in use in Sweden today and this concept has proven to be an effective and inexpensive solution to mitigate the release of pesticides to the environment.

Keywords: biobeds, pesticides, peroxidases, white-rot fungi

1. Introduction

It is known that the unsatisfactory management of pesticides and chemicals in general can result in residues in surface and ground waters as well as in large areas of soils. Danish (Helweg, 1994, Spliid, et al., 1999, Stenvang and Helweg,

M.D. Annable et al. (eds.), Methods and Techniques for Cleaning-up Contaminated Sites, 145–151.

2000), German (Fischer, et al., 1998a, Fischer, et al., 1998b, Frede, et al., 1998, Seel, et al., 1966) and Swedish (Kreuger, 1999) experiences have shown that point sources of pesticides are one of the most dominant reasons for pesticide pollution of creeks, streams and lakes, groundwater and local water supplies.

An important point source of contamination is the filling or cleaning of spraying equipment. This activity is often done at the same location on the farms due to the convenience of a water supply. High concentrations of pesticide residues have been found at such sites (Helweg, 1994). If the spill takes place in a farmyard where the topsoil layer has been replaced by a layer of gravel and sand, there is an obvious risk of groundwater contamination from leaching.

However, the use of biobeds has minimized the risks of pollution when filling the spraying equipment.

2. What is a Biobed?

A biobed is a simple and inexpensive construction on farms intended to collect and degrade spills of pesticides (Torstensson, 2000, Torstensson and Castillo, 1997). Biobeds are facilities composed of a biomixture of straw, mineral topsoil and peat. A grass layer on the top and a clay layer at the bottom are also part of the biobed system. A driving ramp is necessary for the parking of the spraying equipment (Fig. 1).

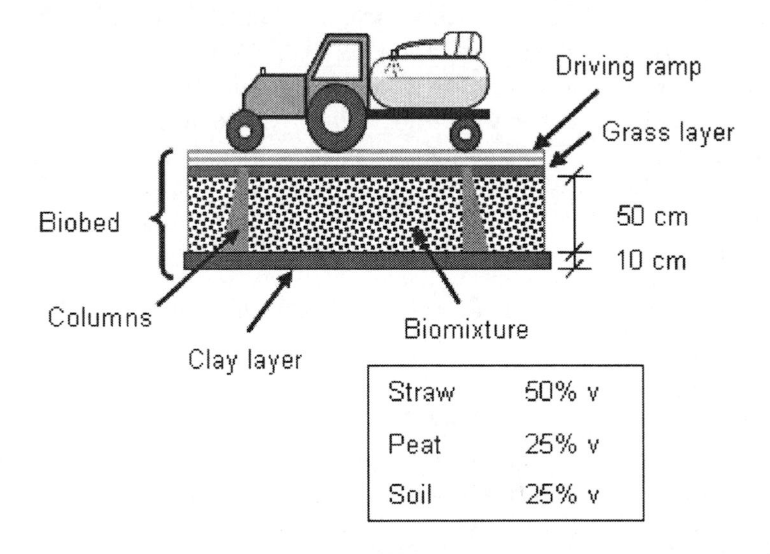

Figure 1. Diagram of a biobed.

The idea behind the biobed is that all handling of pesticides when filling the spraying equipment should be done above the biobed so when spills occur they will be retained and degraded in the biobed. The composition of the biomixture is intended to sustain these activities.

The typical Swedish biomix consists of straw, topsoil and peat (50-25-25 % v/v). The soil provides sorption capacity and degrading microorganisms and the peat contributes to high sorption capacity and to regulate the humidity of the system. A grass layer covering the biobed also helps to maintain the correct humidity and can be used as an indicator to reveal pesticide spills (Fig. 2). The main microbial activity producing degradation of the pesticides comes from straw degradation. The straw stimulates the growth of lignin-degrading fungi and the activity of ligninolytic enzymes (such as manganese and lignin peroxidases and laccases), which can degrade many different pesticides (Castillo, 1997, Castillo, et al., 1997, Castillo, et al., 2000, Castillo, et al., 2001, von Wirén-Lehr, et al., 2001).

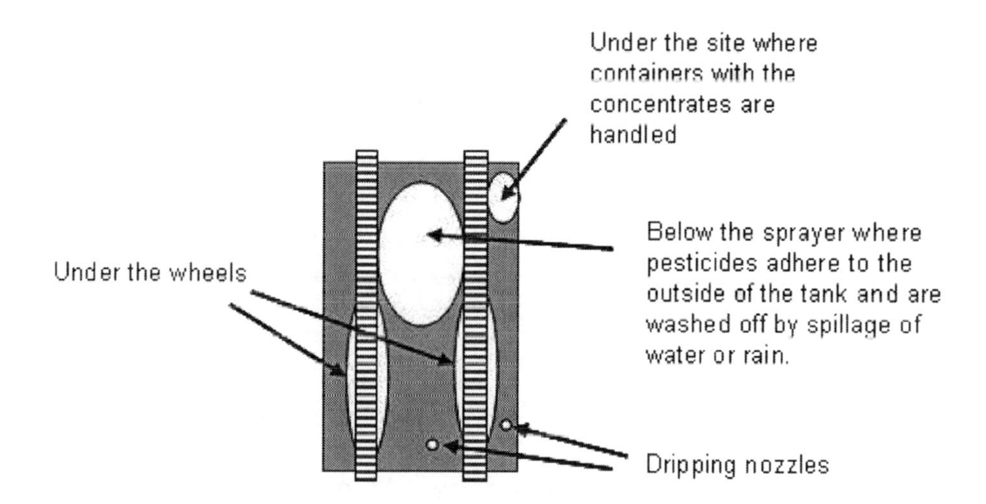

Figure 2. Spillage pattern in a biobed (modified from Torstensson, 2000).

The period where the highest pesticide levels are observed in a biobed is during the spraying season, i.e. when they are used more intensively (Torstensson, 2000, Torstensson and Castillo, 1997). Once spilled, the pesticides are retained in the upper part of the biobed and most of them are degraded within one year (Fig. 3). Levels near or below the detection limit are found in the lower levels of the biomixture suggesting low transport to the bottom.

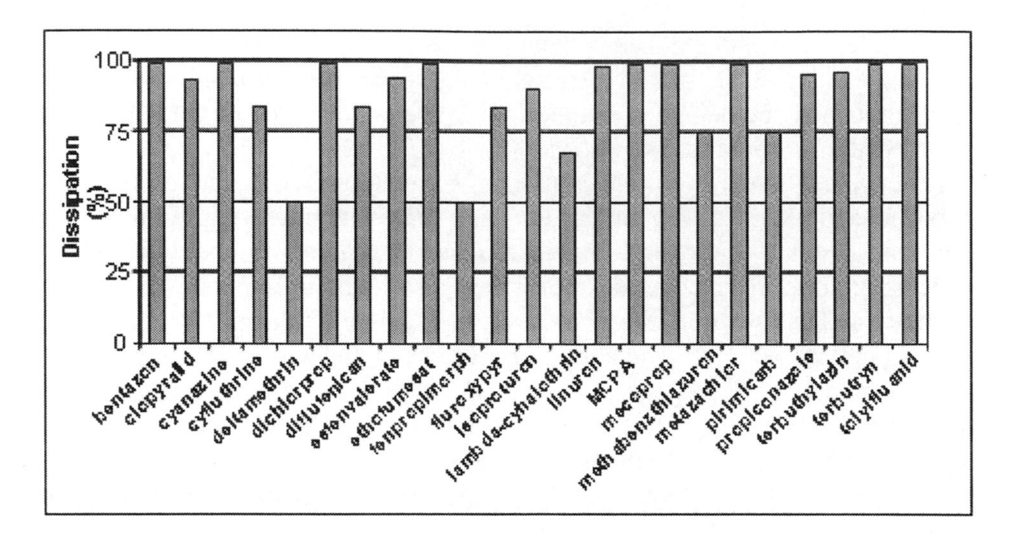

Figure 3. Dissipation of pesticides in biobeds after one year.

Due to degradation of the straw, the carbon content in the core of the biomixture decreases with time to levels similar as the ones in agricultural soils. Therefore, the biomixture should regularly be replaced with a fresh one. This is recommended to be done every 6 to 8 years under Swedish conditions. The removed material may contain small amounts of pesticides residues, either from pesticides used just before removing the biomixture or from pesticides that are slowly degraded. Therefore, we recommend an after-composting process of one year to decrease the levels of the pesticide residues to below the limit of detection (Table 1).

TABLE 1. An after-composting of the removed biomixture reduces the levels of the pesticides residues to levels under the limit of detection. Data from a biobed at the south of Sweden.

Pesticide	April	August	Oct	Dec	Limit of detection
Difuflenican	0.30	0.07	<0.05	<0.05	0.05
Esfenvalerate	0.16	0.11	0.06	<0.02	0.02
Fenpropimorph	0.10	0.04	<0.04	<0.04	0.04
Isoproturon	0.07	<0.01	<0.01	<0.01	0.01
Metazachlor	0.08	<0.04	<0.04	<0.04	0.04
Menthabenzthiazuron	0.10	<0.05	<0.05	<0.05	0.05
Pirimicarb	0.07	0.03	<0.02	<0.02	0.02
Propiconazole	0.12	0.06	<0.05	<0.05	0.05
Terbuthylazine	0.11	0.08	0.04	<0.04	0.04

Due to the Swedish climate the activity in the biobed is limited to the spring, summer and part of the autumn. The highest temperatures are observed during the summer and can reach 20°C in biobeds at the south of Sweden. During the winter the temperatures can be 2 - 4°C (Fig. 4) and low or no microbial activity is expected. Lower temperatures are observed in biobeds in the north of Sweden (data not shown) especially in winter where the middle part of the biobed freezes.

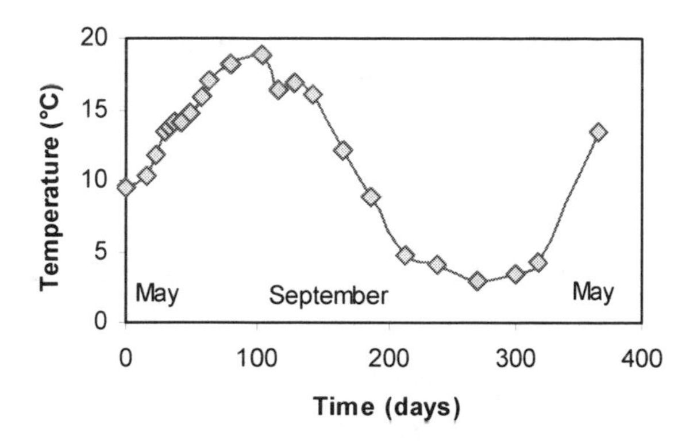

Figure 4. Temperature in a biobed during a period of one year. The biobed is located at the south of Sweden.

The water balance is a delicate issue in a biobed. Non-saturated conditions are needed to allow aerobic degradation processes which are important for most pesticides. Fewer pesticides are degraded under near saturated conditions, which also increase the risk for transport and leakage of pesticides from the biobed. To avoid this situation the Swedish biobeds are not recommended for treatment of large amounts of waters, for instance those from internal washing of the sprayer. This activity should be done at the fields. The moisture in the Swedish biobeds is kept at an optimal level for microbial activity if they are subjected to normal rainfall and to the water coming from the external washing of the sprayer (Fig. 5).

The first biobeds were built in 1993 and it is estimated that there are more than 1500 biobeds in Sweden today. This successful spread is due to the fact that the biobed is simple, effective and inexpensive and therefore has gained acceptance among the Swedish farmers who in turn have developed several models by using materials at hand from the farms (Fig. 6).

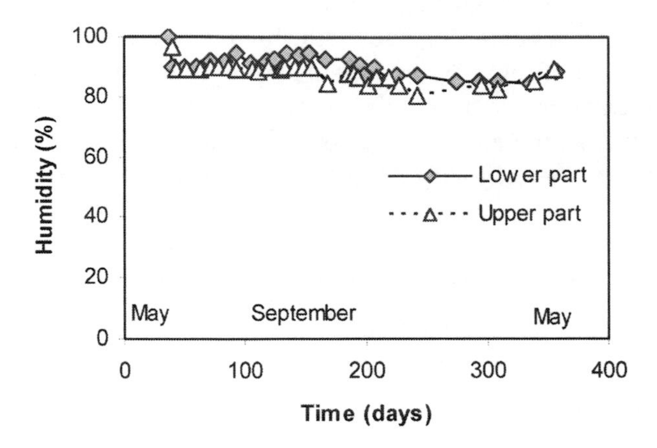

Figure 5. Relative humitidy in a biobed during a period of one year. The biobed is located at the south of Sweden.

Figure 6. Some examples of biobed models, a) with a wooden driving ramp and b) with an old iron lattice.

3. Biobeds in Other Countries

The biobed system has some interesting properties (e.g. effective and cheap) thus it has gained interest in many other European countries, for example Denmark, Finland, Norway, France and Great Britain. The system also is of interest in developing countries since it does not require much maintenance or skilful expertise. Biobeds today are being introduced in Latin American countries such as Peru and Guatemala.

4. References

Castillo, M.d.P. (1997) Degradation of pesticides by *Phanerochaete chrysosporium* in Solid Substrate Fermentation. Doctoral Thesis, Swedish University of Agricultural Sciences.

Castillo, M.d.P., Ander, P. & Stenström, J. (1997) Lignin and manganese peroxidase activity in extracts from straw solid substrate fermentations. *Biotechnology Techniques* 11, 701-706.

Castillo, M.d.P., Ander, P., Stenström, J. & Torstensson, L. (2000) Degradation of the herbicide bentazon as related to enzyme production by *Phanerochaete chrysosporium* in a solid substrate fermentation system. *World Journal of Microbiology & Biotechnology* 16, 289-295.

Castillo, M.d.P., von Wirén-Lehr, S., Scheunert, I. & Torstensson, L. (2001) Degradation of isoproturon by the white rot fungus *Phanerochaete chrysosporium*. *Biology and Fertility of Soils* 33, 521-528.

Fischer, P., Hartmann, H., Bach, M., Burhenne, J., Frede, H.G. & Spiteller, M. (1998a) Gewasserbelastung durch Pflanzenschutzmittel in drei Einzugsgebieten. *Gesunde Pflanzen* 50, 142-147.

Fischer, P., Hartmann, H., Bach, M., Burhenne, J., Frede, H.G. & Spiteller, M. (1998b) Reduktion des Gewasserreintrags von Pflanzensschutzmitteln aus Punktquellen durch Beratung. *Gesunde Pflanzen* 50.

Frede, H.G., Fischer, P. & Bach, M. (1998) Reduction of herbicide contamination in flowing waters. *Zeitschrift Fur Pflanzenernahrung Und Bodenkunde* 161, 395-400.

Helweg, A. (1994) Threats to water quality from pesticides—Case histories from Denmark. *Pesticide Outlook* 5, 12-18.

Kreuger, J. (1999) Pesticides in the environment - atmospheric deposition and transport to surface waters. PhD.

Seel, P., Knepper, T.P., Gabriel, S., Weber, A. & Haber, K. (1966) Kläranlagen als Haupteintragspfad für Pflanzenschutzmittel in ein Fliessgewässer - Bilanzierung der Einträge. *Vom Wasser* 86, 247-262.

Spliid, N.H., Brüsch, W., Jacobsen, O.S. & Hansen, S.U. (1999) Pesticide point sources and dispersion of pesticides from a site previously used for handling of pesticides DJF Report 9.

Stenvang, L. & Helweg, A. (2000) Minimizing pollution risk at filling and washing sites for sprayers DJF report 23.

Torstensson, L. (2000) Experiences of biobeds in practical use in Sweden. *Pesticide Outlook* 11, 169-212.

Torstensson, L. & Castillo, M.d.P. (1997) Use of biobeds in Sweden to minimize environmental spillages from agricultural spraying equipment. *Pesticide Outlook* 8, 24-27.

von Wirén-Lehr, S., Castillo, M.d.P., Torstensson, L. & Scheunert, I. (2001) Degradation of isoproturon in biobeds. *Biology and Fertility of Soils* 33, 535-540.

INCD-ECOIND EXPERIENCE IN BIO-TREATMENT OF THE ORGANIC COMPOUNDS AND HEAVY METALS POLLUTED SOILS

ELISABETA PENA LEONTE, MARGARETA NICOLAU, ILEANA GHITA, RADU MITRAN, VIRGIL ROSCA, CIPRIAN DUMITRESCU AND COSTEL BUMBAC [1]

[1]*National Research & Development Institute for Industrial Ecology – ECOIND; 90-92 Panduri street, sector 5, zip 050663, telephone 04.021/410.03.77, fax: 04.021.410.05.75, e-mail: ecoind@incdecoind.ro; web site: www.incdecoind.ro, Bucharest, Romania*

Abstract. Nowadays, in Romania, soil management and assurance of soil quality conditions are some of the major tasks, not only in respect to reducing soil's value in agriculture and its utilization in recreative or habitat purposes but also because it is a source of pollution for ground and surface water.

Among the conventional technologies of soil decontamination applied in Romania, biotechnological methods are the most important. INCD – ECOIND, one of Romanian National Institutes for the Environmental Protection has developed since 2000 a series of biotechnologies for treatment of soils polluted with oil products, cyanide and heavy metals: I. Biotechnologies in bioreactors inoculated with specialized bacteria; II. Biotechnologies for cyanide polluted soils decontamination (soil was inoculated with autochthon/selected biomass and supplemented with nutrients); III. Treatments with natural sorbents (peat-moss; *Sphagnum sp.*); IV. Phytoremediation using *Dactylis glomerata. Lolium perene, Avena sativa, Hordeum vulgare* for heavy metals removal from contaminated soils. The best results were obtained with *Dactylis glomerata.*

Choosing the best bioremediation technology must be made on the basis of technical and economical criteria, which take into account the following: type and concentration of present pollutants, polluted soil volume which need to be treated, its localization (depth) and accessibility, treatment objectives or legal regulations;

M.D. Annable et al. (eds.), Methods and Techniques for Cleaning-up Contaminated Sites, 153–163.

The technologies elaborated by INCD-ECOIND are efficient as the treated soils are conformed to Romanian regulation – MAPPM Order 756/1997 "Reglementation regarding evaluation of environment contamination".

Keywords: soil decontamination; oil products, heavy metals, cyanide, phytoremediation

1. Introduction

The role of soil is of vital importance to humankind and the maintenance of a healthy natural environment. The 1972 European Soil Charter recognized for the first time that any biological, physical or chemical degradation of soil should therefore be of primary concern and that appropriate measures to protect soils should be implemented without delay.

Past and present economic activities have often resulted in contamination of the soil underlying the place where these activities took place. Contaminants affecting soil may be in the form of solids, liquids or gases. The most common toxic soil pollutants include metallic elements and their compounds, organic chemicals, oils and tars, pesticides, explosive and toxic gases, radioactive materials, biologically active materials, combustible materials, asbestos and other hazardous minerals. These substances commonly arise from the disposal of industrial and domestic waste products in designated landfills or uncontrolled dumps (Van Lynden G W J, 1994).

Various technologies are available for treatment (Neag Gh., 1997) These include soil excavation, washing and disposal. However, this type of treatment is very expensive, cheaper technological solutions being required. Among the technologies of soil decontamination applied in Romania, biotechnological methods are the most important. Bioremediation technologies of soils polluted with oil compounds, heavy metals and cyanides have a huge application potential as they are cost-effective and can be applied in situ.

This paper presents some results for:

- oil polluted soils decontamination using specialized bacteria,
- biotechnologies for cyanide polluted soil decontamination,
- remediation of soils contaminated with oil products using biodegradable natural sorbents,
- phytoremediation technologies using *Dactylis glomerata, Lolium perene, Avena sativa, Hordeum vulgare* for heavy metals removal from contaminated soils.

1.1. BIOTECHNOLOGIES IN BIOREACTORS WITH ADDITION OF SPECIALIZED BACTERIA AND NUTRIENTS FOR OIL POLLUTED SOIL DECONTAMINATION

Enhanced remediation of crude oil polluted soil can be achieved by accelerating the biodegradation process through seeding the bioreactors with adapted microorganisms and nutrient addition (Ponchon, R. et al., 1995)

The laboratory experiments were conducted in three aerobic batch bioreactors as follows:

- P III N - inoculum (autochthonous biomass) with N and P nutrients addition in a 100C:10N:1P ratio;
- P III A – inoculum (selected by adaptation biomass) with nutrient addition;
- P III M – inoculum (autochthonous biomass) without nutrient addition;

Initial oil product concentration in soil samples was 2100 mg/kg d.w., which exceeded by far the oil products concentration in soils normed by the 756/1997 Ordinance (100 mg/kg d.w.);
Experiment duration – 40 days;

1.1.1. *Experimental Results*

Oil product concentrations and removal efficiencies were measured after 7, 14, 28 and 40 days. Removal efficiencies obtained varied between 56 to 98 percent. Experimental results are presented in table no. 1 and figure 1.

TABLE 1. Oil products removal in biodegradation process in experimental batch bioreactors

Experimental variants	Oil products concentration mg/kg d.w	HRT = 7 d		HRT = 14 d		HRT = 21d		HRT = 28d		HRT = 40 d		Global efficiency, %
		o.p. mg/kg d.w	η %	o.p. mg/kg d.w	η %	o.p. mg/kg d.w	η %	o.p. mg/kg d.w	η %	o.p. mg/kg d.w	η %	
P III N	2100	1140	45	480	72	130	73	82	37	-	-	96
P III A	2100	670	68	182	72	30	83	-	-	-	-	98
P III M	2100	1660	21	1200	27	1100	8	923	16	923	-	56

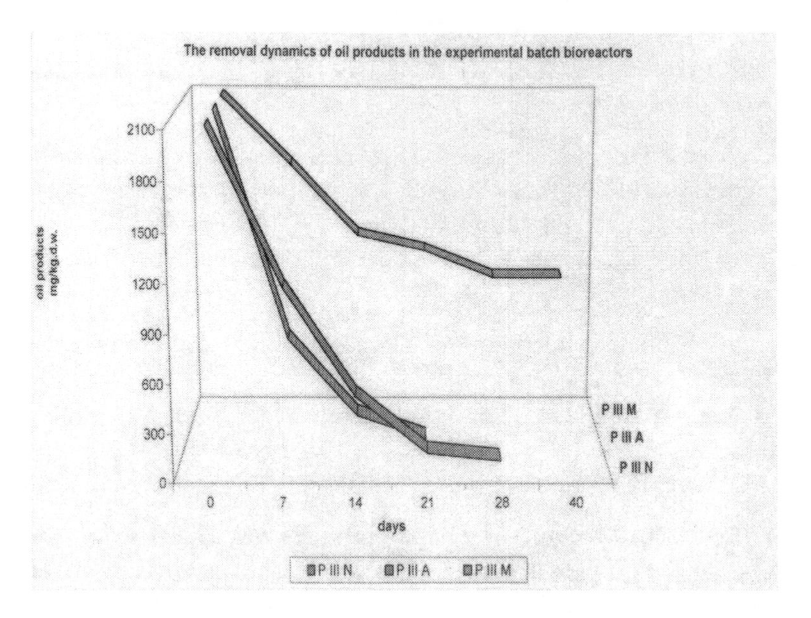

Figure 1. Removal dynamics of oil products in the experimental batch reactors

Identified bacterial types and theirs percentage frequency in unselected and selected bacterial inoculum is emphasized in table no. 2. The dominant bacterial type is *Arthrobacter,* 40% in bioreactor P III N and 60% in bioreactor P III A.

TABLE 2. Bacterial types and theirs frequency in unselected and selected bacterial inoculum

Bioreactor P III N unselected biomass	**Bioreactor P III A selected biomass**
Arthrobacter 40%	Arthrobacter 60%
Bacillus 20%	Bacillus 30%
Pseudomonas 10%	Pseudomonas
Flavobacterium < 5%	Agrobacterium
Micrococcus < 5%	Micrococcus
Mycobacterium < 5%	Flavobacterium (Pseudomonas, Agrobacterium, Micrococcus, Flavobacterium: 10%)
Agrobacterium < 5%	
Azotobacter < 5%	

- The microorganisms supply in the experimental bioreactor lead to:
  - Diminution of the retention time to 21 day;
  - Increase of the oil products removal efficiencies from 95 to 98% (P III N and P III A);
- Lack of nutrients lead to the decrease of the degrading bacterial micro-organisms number and consequently to a decrease of the bioremediation efficiency to 56% (P III M).

1.2. BIOTECHNOLOGIES FOR CYANIDE POLLUTED SOILS DECONTAMINATION

Soil samples (S1) to be treated were withdrawn from an area located near the coke-chemical plant UCC 2 – at SIDEX Galati. The cyanide content of the samples was enriched with up to 208 mg/kg d.w. by zinc-cyanide addition.

The experiments were conducted in two variants:

- Variant I – with soil natural seeded with autochthon microorganisms;
- Variant II – with soil enriched with activated sludge microorganisms.

In both experimental variants was used the "bulk method":

In conformance with experimental procedure:

- polluted soil was stored on an impermeable surface and maintained with a humidity of about 30%;
- nitrogen and phosphorus were added to assure a ratio between organic compounds and nutrients of 100C:10N:1P;

The experimental period was of 40 days.

1.2.1. *Experimental Results*

Cyanide concentrations for all experimental variants were measured after 28, 35 and 39 days. Experimental results are presented in table no. 3 and 4.

TABLE 3. Experimental results for variant I – soil with autochthon microorganisms

Crt. no.	Parameters/days of experiment	Samples S1			
		0	28	35	39
1	pH	8,59	8,42	8,36	8,47
2	CN- mg/kg d.w.	208,5	39,3	30,1	27,6

Efficiencies of cyanide removal for variant I varied between 81% (after 28 days) and 86% (after 39 days).

TABLE 4. Experimental results for variant II – soil enriched with activated sludge micro-organisms.

Crt. no.	Parameters/days of experiment	Samples S1			
		0	28	35	39
1	pH	8,59	8,42	8,36	8,47
2	CN- mg/kg d.w.	208,5	37,9	30,9	20,9

Efficiencies of cyanide removal for variant II varied between 82% (after 28 days) and 90% (after 39 days).

In conclusion, autochthon microflora can remove cyanides with high efficiencies, in a short period of time (≈ one month) from soils polluted with relative high concentration – 200 mg/kg d.w. Very little difference was observed on cyanide removal between the two experimental variants.

After 40 days of experiment, the remaining cyanide concentration was about 20 mg/kg d.w., which is under the alert limit normed by 756/1997 Ordinance of the Water and Environmental Protection Ministry.

1.3. REMEDIATION OF SOILS CONTAMINATED WITH OIL PRODUCTS USING BIODEGRADABLE NATURAL SORBENTS

As biosorbent was used peat moss (*Sphagnum sp.*) in comparison with the commercial product "SpillSorb" and were established the sorbent capacity and doze of sorbent per kg of polluted soil.

For this experiment, four experimental variants were conducted:

- Variant A - 0,4 g SpillSorb added per kg of sandy soil polluted with oil products in concentration of 825 mg/kg d.w.;
- Variant B - 0,4 g of peat moss (*Sphagnum sp.*) added per kg of sandy soil polluted with oil products in concentration of 887 mg/kg d.w.;
- Variant C - 0,4 g SpillSorb added per kg of loamy soil polluted with oil products in concentration of 998 mg/kg d.w.;
- Variant D - 0,4 g of peat moss (*Sphagnum sp.*) added per kg of loamy soil polluted with oil products in concentration of 1098 mg/kg d.w.;

Experimental conditions:

- Aerobe batch bioreactors;
- Substrate composition: 1000 g of polluted soil, 200 ml of distilled water, 0,4 g of peat moss ((*Sphagnum sp.*) or SpillSorb;
- Temperature of 20-25°C;
- pH – neutral to slightly alkaline;
- Humidity of 65-85%.

1.3.1. *Experimental Results*

Oil products concentrations for all experimental variants were measured after 1, 2, 3, 4, 5, 12 and 19 days. Dynamics of the biodegradation process of oil products in batch bioreactors is emphasized in table no. 5 and 6 and fig. no. 2.

TABLE 5 and 6. Dynamics of the biodegradation process of oil products

Exp. Var.	Oil products concentration determined in sediment mg/kg d.w.											Global efficiency after 5 days %
	Initial	HRT 1 d.	η %	HRT 2 d.	η %	HRT 3 d.	η %	HRT . 4 d.	η %	HRT 5 d.	η %	
A	825	503	39	432	14	313	28	298	5	289	3	65
B	887	729	18	513	30	359	30	324	10	315	3	64
C	998	607	39	604	1	528	13	413	22	410	1	59
D	1098	900	18	686	24	509	26	459	10	426	1	61

Exp. Var.	Oil products concentration determined in sediment mg/kg d.w.					Global efficiency after 19 days %
	Initial	HRT 12 d.	η %	HRT 19 d.	η %	
A	825	263	9	256,3	3	70
B	887	277,9	12	270,3	3	70
C	998	386,7	6	384,14	1	68
D	1098	392,6	8	386,2	2	65

After 19 treatment days were obtained oil products removal efficiencies varying between 65-70% depending on soil texture and pollutant concentration.

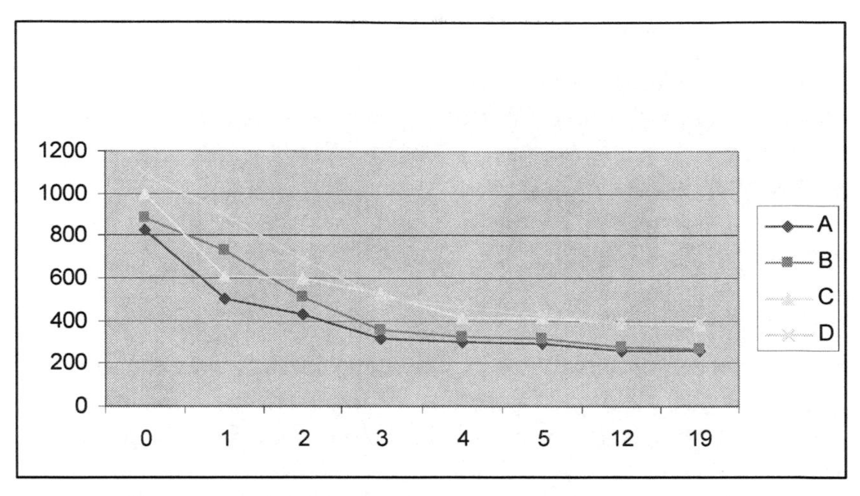

Figure 2. Dynamics of the biodegradation process of oil products batch bioreactors

1.4. PHYTOREMEDIATION OF HEAVY METALS POLLUTED SOILS

Bioavailability of heavy metals in metal-contaminated soils depends on physical, chemical and biological factors. Physical (structure, penetrability) and chemical factors (pH, speciation, and concentration) give the framework in which biological factors can modify the metal availability by release of oxygen, protons, and organic acids and by association with micorhizal fungi. With these conditions in mind, the possibilities of the use of hyper accumulating higher plants for the decontamination of metal polluted soils, i.e. phytoremediation, are explored (Ernst, W.H.O,1996).

For the experiments, we used soil samples withdrawn from a mine located near Certez – Maramures. Soil samples were polluted with Mn, Zn and Pb ions. The initial concentrations of heavy metal ions in soil samples were: Mn = 3362 mg/kg d.w.; Zn = 815 mg/kg d.w.; Pb = 508 mg/kg d.w.

The green plants used as biosorbents were: *Hordeum vulgare* (two-row barley), *Lolium perene* (english raigras – dornel), *Dactylis glomerata* (orchard grass), *Avena sativa* (oat).

The green plants were seeded in spring and harvested in autumn and the heavy metal ions were determined from whole plants after harvesting.

1.4.1. *Experimental Results*

Heavy metal concentrations removed from soil for all experimental variants are emphasized in fig. no. 3, 4 and 5.

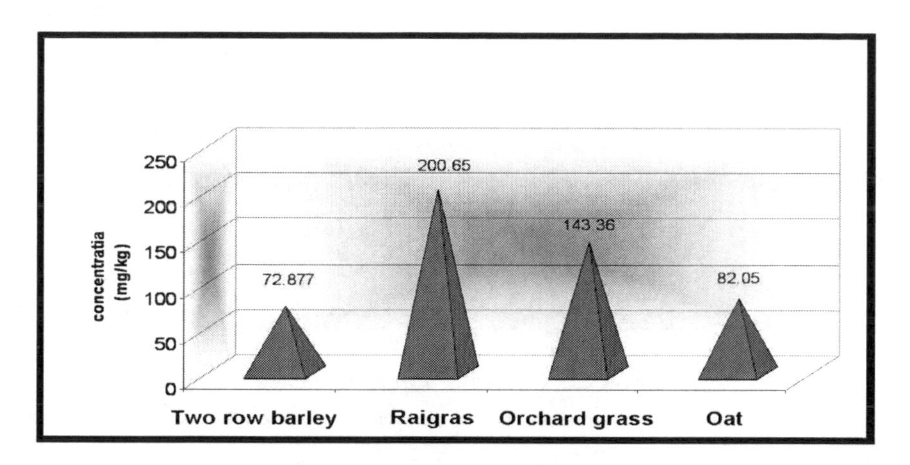

Figure 3. Mn ions concentrations in plants

The maximum Mn ions concentration was detected in *Loliul perene* – english raigras (about 200 mg/kg d.w.).

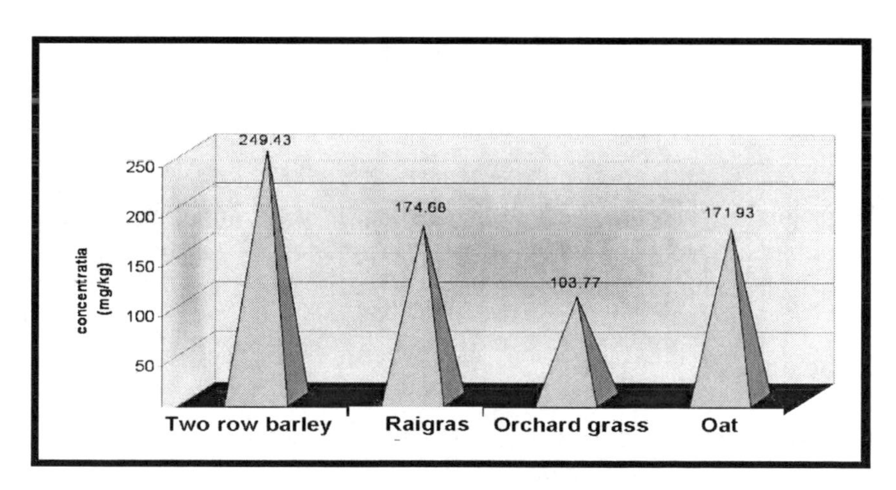

Figure 4. Zn ions concentrations in plants

The maximum Zn ions concentration was detected in *Hordeum vulgare* – two row barley (about 250 mg/kg d.w.).

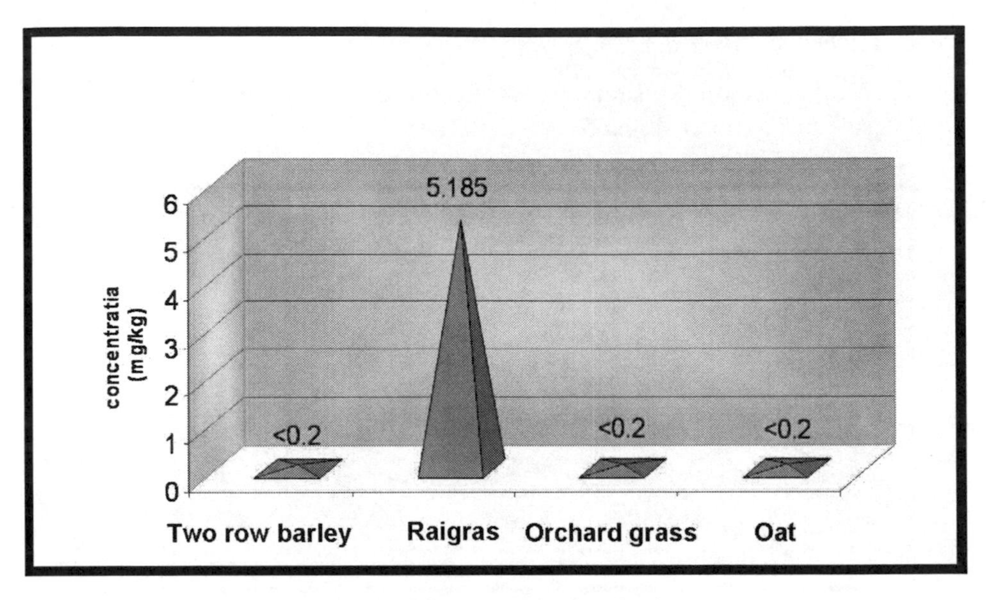

Figure 5. Pb ions concentrations in plants

Pb ions were removed by green plants with little efficiency, probably because of the reduced availability from soil.

Results and discussions:

- The experimental results will be continued for the elaboration of a culture technology for ecological rehabilitation of areas polluted with heavy metal ions;
- The major problem which remains to be solved by future researches is the ecological disposal of the contaminated plants;

Biotechnological researches in this field are of great importance because in Romania, mine closing left behind large areas polluted with heavy metals which need to be rehabilitated.

2. Conclusions

In our institute were obtained, in laboratory conditions, some promising preliminary results:

- 98% global removal efficiency of oil products after 40 days using a experimental batch bioreactor inoculated with adapted microorganisms and with nutrient addition;

- 82% (after 28 days) and 90% (after 39 days) removal efficiencies of cyanides from soils enriched with activated sludge microorganisms as inoculum;
- Oil products removal efficiencies varying between 65-70% using peat moss (*Sphagnum sp.)* as biosorbent, depending on soil texture and pollutant concentration
- Phytoremediation of heavy metal polluted soils:
  - The maximum Mn ions concentration was detected in *Loliul perene* – english raigras (about 200 mg/kg d.w.).
  - The maximum Zn ions concentration was detected in *Hordeum vulgare* – two row barley (about 250 mg/kg d.w.).
  - Pb ions were removed by green plants with little efficiency, probably because of the reduced availability from soil.

Choosing the best bioremediation technology must be made on the basis of technical and economical criteria, which take into account the following: type and concentration of present pollutants; polluted soil volume which need to be treated, its localization (depth) and accessibility; treatment objectives or legal regulations;

References

Ernst, W.H.O. (1996). Bioavailability of heavy metals and decontamination of soils by plants. Applied Geochemistry, **11**, 163–167.

Neag Gheorghe, (1997). Soils and groundwater decontamination, Casa Cartii de Stiinta, ISBN: 973-9204-70-8.

Ponchon, R., Muth, R., Bias, J.J. (1995). *La biodegradation des hydrocarbures aromatiques polycycliques*. L'eau, l'industrie, les nuissances, **187**, 45–49.

Van Lynden, G.W.J. (1994). The European soil resource: current status of soil degradation in Europe: causes, impacts and need for action. ISRIC, Wageningen. Council of Europe, Strasbourg.

REVIEW OF CHARACTERIZATION AND REMEDIATION TECHNOLOGIES FOR NAPL'S IN GROUNDWATER

WALTER W. KOVALICK, JR.*
Assistant Regional Administrator for Resources Management
US Environmental Protection Agency, Chicago, Illinois

1. Introduction

The principal legislation dealing with contaminated sites in the United States was enacted in 1980. It dealt with abandoned hazardous waste dumps; since then additional legislation has been passed to deal with corrective action from releases at operating industrial facilities, leaking underground gasoline storage tanks, and redevelopment and renewal of central urban areas thought to be contaminated (i.e. the restoration of Brown fields). As each of these programs has been implemented, the engineering community has worked with technology developers to utilize new technologies to solve these problems. This paper describes some of the major techniques and approaches that have evolved over the last 25 years and the information resources that are now available to those in other countries facing these contaminated soil and groundwater problems for the first time. This paper discusses the current technologies being used, primarily with data from the U.S. Superfund program, and then, with particular emphasis on the detection and clean up of dense non-aqueous phase liquids (DNAPL's), reviews the latest groundwater monitoring and assessment approaches as well as remediation technologies. Finally, there is a brief discussion of new and emerging approaches being explored in the U.S.

2. U.S. Superfund Program Technology Data

For groundwater cleanup, Figure 1 shows variety of approaches being taken at 849 Superfund sites for the period from 1980 through 2004. While 712 sites are using traditional pumping to the surface and treating in fixed facilities, there are

* Walter W. Kovalick Jr., Ph.D., Assistant Regional Administrator for Resources Management US Environmental Protection Agency, Chicago, Illinois

M.D. Annable et al. (eds.), Methods and Techniques for Cleaning-up Contaminated Sites, 165–175.

a growing number of sites (primarily in the last 8-10 years) using alternative approaches or other approaches coupled with pump and treat. While 24 sites are using some type of in situ treatment technology alone, a total of 133 sites are using such in situ treatment coupled with other techniques—including pump and treat and monitored natural attenuation.

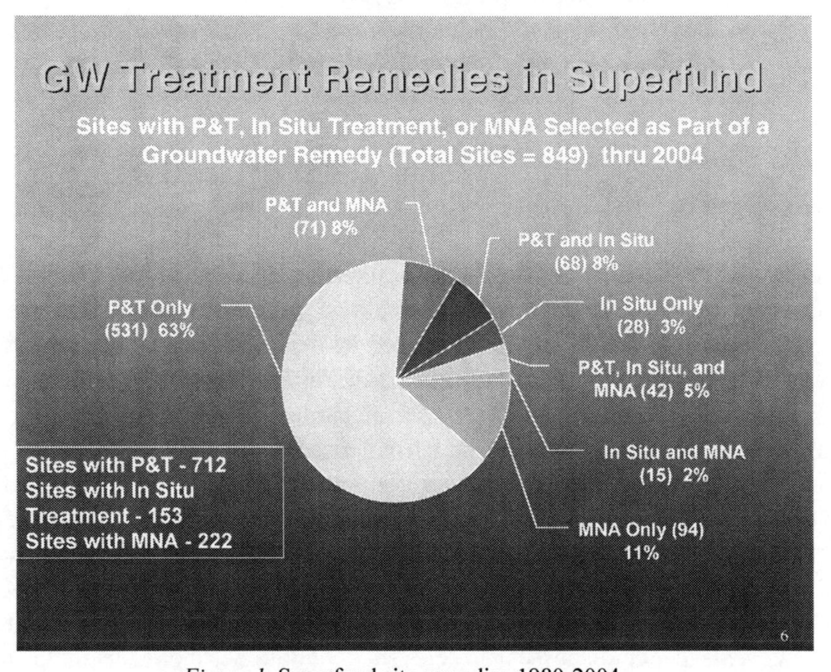

Figure 1. Superfund site remedies 1980-2004

EPA (in cooperation with the states in the U.S.) have developed a precise definition of "monitored natural attenuation", which involves developing lines of evidence with a monitoring program to assure that pollution is actually being mitigated. MNA is being used alone at 94 sites, while it is part of a remedy "train" at 222 sites. MNA is even more widespread as a potential remedy at U.S. leaking underground storage tank sites, at which petroleum hydrocarbon is the principal contaminant vs. Superfund sites which have many other organic compounds. More information on the use of these technologies in the Superfund program can be found in the EPA Annual Status Report at http://cluin.org/asr.

Technology information related to contaminated soils is also found in this report.

3. Site Characterization Technology Developments

One of the most troubling aspects of clean up of chlorinated solvents (one of the most common of the DNAPL's) is precisely locating them. This is because they are denser than water and tend to sink and follow the heterogeneities in the subsurface. They form ganglia in the groundwater, so that locating the source term as well as the downstream chemicals is very difficult—even before they begin to diffuse in the groundwater.

As a basic primer on field analytical technologies for site investigation, EPA in cooperation with the Army Corps of Engineers has developed FATE—the Field Analytical Technologies Encyclopedia. An on-line tool, FATE is designed to allow the user to choose a contaminant or a technology and then "drill down" for information about its operation, use, contaminant sensitivity, etc. It covers 17 technology classes including:

- Chemical analytics (e.g. fiber optic chemical sensors, gas chromatography, graphite furnace AA spectroscopy, immunoassay, infrared spectroscopy, laser-induced fluorescence, test kits, x-ray fluorescence)
- Direct push technologies (analytical systems, direct push platforms, geo-technical sensors, groundwater samplers, soil & soil-gas samplers)
- Geophysics (ground penetrating radar, magnetics)
- Comprehensive explosives module

FATE can be found at www.cluin.org.

Beyond this primer, EPA has developed a more focused discussion of technologies available to help locate DNAPL's in the sub-surface. Site Characterization Technologies for DNAPL Investigations Report reviews these several approaches; while no single technology has been shown to be effective by itself, the report highlights how these several tools that can be combined (using a lines-of-evidence approach) to narrow the location of the DNAPL's. The techniques described and analyzed include:

- Geophysical techniques
  - e.g., acoustic, electromagnetic, electrical, GPR, seismic
- Non-geophysical
  - e.g., diffusion sampler, direct push platform (conductivity, induced fluorescence), in situ groundwater sampling, membrane interface probe, hydrophobic dye, tracer testing, soil gas

The complete report is available at www.cluin.org/download/char/542r04017.pdf What recent experience has shown is that these technologies coupled with real time visualization software can give much more accurate representations of the subsurface in much less time with a much higher level of confidence. This higher level of confidence is due to the hundreds of additional data points that can be used by the visualization software vs. the relatively few resulting from sampling expensive groundwater wells.

In addition to these tools and software developments, EPA and many other public and private entities in the U.S. are using much "smarter" approaches to the planning and conduct of site investigation studies as well as to the monitoring of site clean up systems.

Called the Triad (because of its three parts), this approach brings together systematic planning, dynamic work planning, and real time data under one umbrella to optimize the taking of data and the anticipation of its uses. The approach is designed to fully maximize the capabilities of field analytical instruments and rapid sampling tools.

Systematic planning involves the identification of project-specific goals or desired data sets vs. the use of prescriptive methods "checklists" or following standard grids for taking samples. It relies on thorough advance planning and an up-front understanding of the site. Such planning takes a global view of project, and its ultimate goals vs. just the needs of the design engineer, the risk assessor, the regulator, or any other single client. The whole purpose is to gather data to manage the uncertainty of decisions to the appropriate level. The dynamic work planning or adaptive decision making allows the site team to anticipate uncertainties and developments while in the field with real time instruments and develop decision rules on what additional sampling should be avoided or should be conducted to reduce the uncertainty of future decisions. This avoids multiple mobilizations on the site, thus reducing costs. This entire approach is made possible by the advent of the field analytical instruments and the ability to do real time data acquisition and analysis that has evolved in the last 10 years. The tools described in FATE have made this approach feasible.

Figure 2 is an illustration of the nature of the improved decision making that can result from the use of the Triad approach. The left side of the diagram shows the taking of a few expensive, traditionally grided samples in an attempt to define site contamination—as indicated by ellipse and the other geometric shape. These few high quality laboratory data points yield a low level of information in terms of defining these shapes. The right side of the diagram show a much more intensive use of field analytical techniques that may be of lower precision in terms of contaminant concentration, but yield much more information to reduce uncertainty in locating the shapes under the surface.

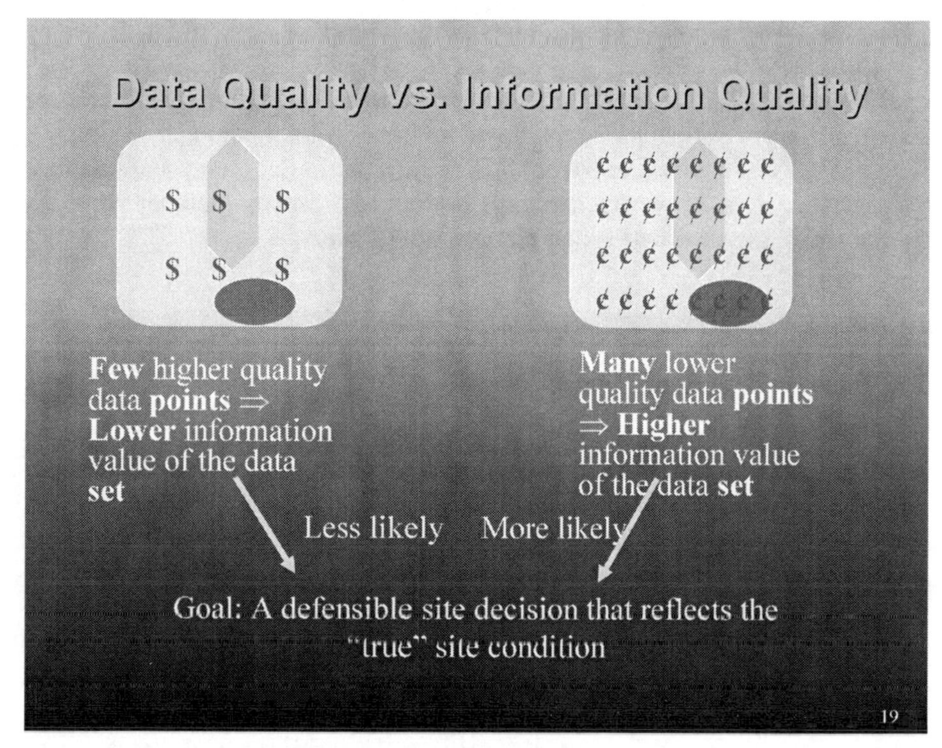

Figure 2. Improved decision making

Related to the software and tools available to support the Triad approach (and for other uses), EPA compiled a list of decision support tools (DSTs) based on information from its laboratories and regions, as well as from other federal agencies including the Department of Energy and the Department of Defense. EPA also developed a set of criteria to use in evaluating these DSTs. These criteria included that the end user was a technically proficient field person such as field project manager – someone able to use a computer, but not a computer modeling expert. The tool had to be a Decision Support tool. The default output should be predictive (decision support) from input. The tool must be freely available to the public. The results of this effort are found at: http://clu-in.org/products/dst

Much work has been done to describe this Triad approach in more detail, offer case examples of its use, examples of contract documents to conduct this type of work, etc. A joint EPA, U.S. Dept. of Energy, and Corps of Engineers web site has been set up at www.triadcentral.org.

4. In Situ Groundwater Treatment Developments

Developments in groundwater treatment technologies are easily divided into two categories: dissolved phase treatment and in-situ treatment technologies.

For dissolved phase treatment, this review assumes pump and treatment at the surface is a known technology. Newer approaches include: permeable reactive walls, bioremediation and phytoremediation.

4.1. DISSOLVED PHASE TREATMENT

4.1.1. *Permeable Reactive Walls*

Permeable reactive walls consist of treatment media placed in the path of the plume allowing passive interception and treatment of the contaminant. Pioneered originally with granular iron placed to treat chlorinated solvents, there are over 200 such walls known world wide to deal with these contaminants.

Newer designs are being experimented with to go deeper into the subsurface and be more efficient. For example, hydro-fracturing is being used in denser media to create treatment zones in which to place the treatment material. Columns or panels are being used as opposed to just treatment walls.

New materials are being for the walls, such as bi-metals for treatment of organics. For inorganic contaminants, a host of new materials are being tried including: apatite, compost, sulfate reducing bacteria, limestone, and other absorptive media. In addition, other forms of zero valent iron are being experimented with including: granular (emplaced walls), micro (injected) and nano scale.

4.1.2. *Bioremediation*

Bioremediation of contaminants in groundwater involves the in situ engineered design for optimizing conditions for microorganisms to degrade dissolved contaminants to less toxic end products. Anaerobic degradation through reductive dechlorination effectively degrades trichloroethylene to nontoxic ethane. Many commercially-available substrates and, if necessary, mixed cultures have been developed to increase the population of reductive dechlorinators. As such, the technology is becoming routinely used at many chlorinated solvent sites.

4.1.3. *Phytoremediation*

In general, phytoremediation refers broadly to multiple processes that take place within and around plants and trees to reduce and eliminate toxic contaminants. These processes may include sequestration in the root systems, degradation in the root zone, transpiration, destruction, and accumulation within the roots, shoots, trunks, and leaves of plants and trees.

For groundwater remediation, there is particular interest in the use of trees as solar-driven "pump-and-treat systems" to control migration and/or remove chlorinated solvents. In the U.S., poplars and cottonwoods, in particular, extend roots to water table, and are very fast growing. An EPA study of 80 phyto-remediation projects in U.S. found over half of them used trees for hydraulic control and/or contaminant removal from groundwater (Use of Field-Scale Phytotechnology for Chlorinated Solvents, Metals, Explosives and Propellants, and Pesticides at http://www.cluin.org). A major issue for remediation projects has been helping project managers understand the issues of agronomy and soil science involved in phytoremediation of groundwater. EPA in concert with academic and industry partners has developed a new protocol for evaluating the applicability and performance of phytoremediation for chlorinated solvents in groundwater and soil (Evaluation of Phytoremediation for Management of Chlorinated Solvents in Soil and Groundwater at http://www.rtdf.org)

Beyond groundwater remediation, EPA has compiled an inventory of some 79 projects using phytoremediation. Use of Field-Scale Phytotechnology for Chlorinated Solvents, Metals, Explosives and propellants, and Pesticides (http://www.cluin.org) contains information of projects in 31 states. Chlorinated solvents were the most frequently treated contaminants using phytoextraction and hydraulic control as the most common techniques. Projects ranged in size from <0.5 acre up to 1000 acres. The report found a lack of published information on cost and performance as well as lessons learned.

4.2. SOURCE TERM TREATMENT

Many other technologies have arisen in the last 5-10 years for in situ treatment of source terms of groundwater contamination, especially DNAPL,s. While in varying stages of development and commercialization, they hold great promise in dealing with the very difficult problems these contaminants create.

This paper provides only a brief summary of these in situ technologies to help distinguish them from the dissolved phase technologies mentioned above. (Considerable more information can be found at www.cluin.org.)

The five major categories and their more specific technology types are:

- In Situ Thermal
    - Steam Enhanced Extraction
    - Electrical Resistive Heating
    - Thermal Conductive Heating

- In Situ Redox Manipulation
  - Chemical Oxidation (ISCO)
  - Chemical Reduction (ISCR)
- Surfactant/Co-Solvent Flushing
- Bioremediation
- Nanotechnology

4.2.1. *In Situ Thermal Processes*

In situ thermal processes involve the use of steam or electricity to raise the temperature of the subsurface and mobilize the contaminants either in the liquid or vapor phase. A network of wells is then used to remove the contaminants and collect or treat them at the surface. Multipurpose injection and extraction wells take advantage of the area of influence of the steam plume to remove the contaminants. Electrical resistance heating and thermal conductive heating use electrodes inserted vertically in the ground on the diameter of a circle with a central recovery well to remove the contaminants. The circles are arranged in an overlapping array to cover a larger area as shown in Figure 3.

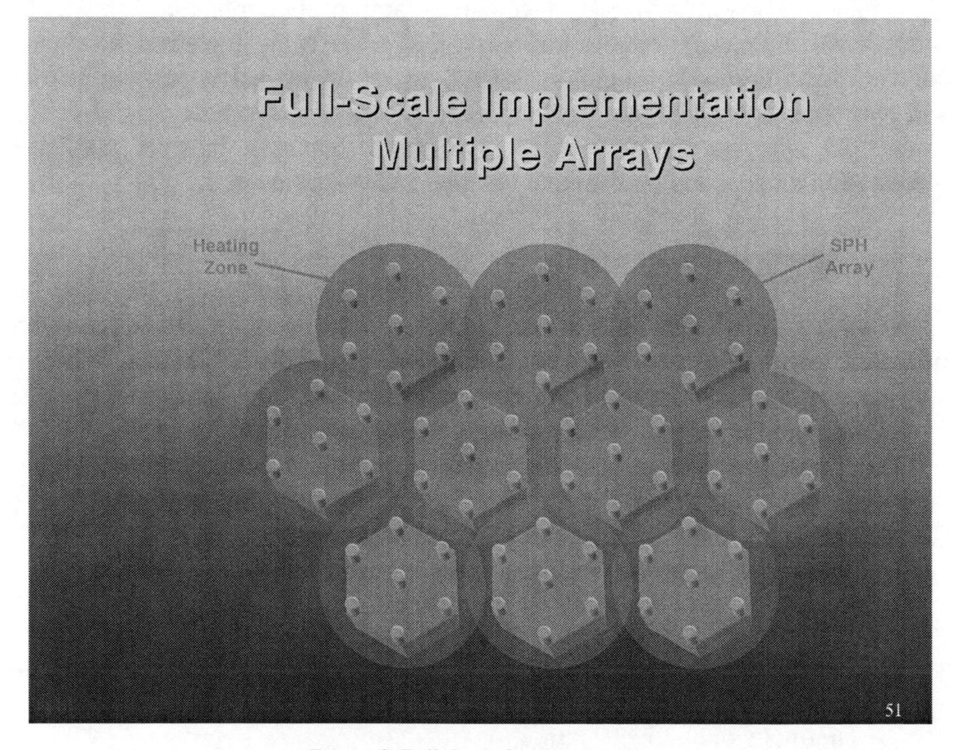

Figure 3. Full-Scale implementation

4.2.2. *In Situ Chemical and Flushing Processes*

In situ chemical reactions have also been developed to oxidize or reduce contaminants in place. Either of these processes can be enhance further biological degradation or be enhanced by such biota. Common in situ oxidants include: Fenton's reagent, hydrogen peroxide, ozone, permanganate (potassium and sodium), and persulfate. Major challenges exist in implementing in situ chemical oxidation remedies. First, it is essential to get contact between oxidants and contaminants; thus, the precision of the delivery mechanism is important. This is particularly challenging in low permeability/heterogeneous subsurface environments. Second, even if the delivery is accomplished, the stoichiometry may be problematic at sites with large quantities of contaminants and/or large amounts of naturally occurring oxidant demand. This condition makes multiple injections much more likely, thus increasing costs.

Surfactant or co-solvent flushing involves the addition of chemicals (such as surfactants and alcohols) to mobilize DNAPL's and remove them through downgradient wells. In addition to the on-going issues of locating the contaminants and their source terms (as noted above), the costs and recovery of the chemicals are of importance in terms of cost of these approaches.

4.2.3. *In Situ Bioremediation*

In situ bioremediation of source terms of DNAPL's is not commercialized, but the subject of international applied research. New data show that microbes grow close to source areas. These microbes (dechlorinators) can increase DNAPL dissolution rates by causing a steeper concentration gradient and by producing more soluble daughter products. In addition, it appears that some substrates added to enhance reductive dechlorination may increase the solubility of the DNAPL. Also, bioremediation is being studied as polishing step for post-treatment residual DNAPL after more aggressive treatment with thermal, chemical, or co-solvent/surfactant flushing. A joint U.S.-U.K. consortium of agencies and companies (in a project named SABRE) is investigating the hypothesis that anaerobic bioremediation using reductive dechlorination can result in the effective treatment of chlorinated solvent DNAPL source areas. Current research status can be found at: http://www.claire.co.uk/sabre_proj.php

4.2.4. *Nanotechnology*

Like bioremediation, nanotechnology is emerging as a new approach to remediation of DNAPL source terms contamination. Field scale research and pilot stage projects are being carried using in situ injection of nanoscale zero-valent iron (NZVI) particles into source areas of groundwater contamination. While the initial contaminants of concern are chlorinated hydrocarbons, metals and pesticides in groundwater are also being studied. EPA has identified over 20 field-scale or full-scale studies underway using NZVI. Further resources on this topic include the issue area at: www.cluin.org/nano as well as the presentations and proceedings from an October 2005 international Workshop on Nanotechnology for Site Remediation held October 20-21, 2005, in Washington, D.C. Proceedings and presentations can be found at: http://www.frtr.gov/nano

EPA and other organizations have developed a number of resources pertaining to these in situ technologies. They include:

- In Situ Thermal Treatment of Chlorinated Solvents: Fundamentals and Field Applications (clu-in.org/techpubs.htm)
- Strategies for Monitoring the Performance of DNAPL Source Zone Remedies (www.itrcweb.org)
- Site Characterization Technologies for DNAPL Investigation (clu-in.org/techpubs.htm)
- DNAPL Remediation: Selected Projects Approaching Regulatory Closure (clu-in.org/techpubs.htm)
- In Situ Thermal Treatment Site Profiles – 90 projects (http://clu-in.org/products/thermal)
- Chemical Oxidation Site Profiles – 389 projects (http://clu-in.org/products/chemox)

In addition to consideration of these individual remedies, more recent developments have embraced a more cost effective and "surgical" approach to clean up, i.e. combined remedies. A combined remedy approach conceptually involves removing, reducing, or destroying the source term of DNAPL's with a concomitant reduction in plume concentration, reach, or other aspects and then the use of further processes (i.e. bioremediation or monitored natural attenuation) to complete the clean up that is required.

5. Conclusion

In large part due to the market need to create better, less costly remediation solutions for contaminated groundwater, dramatic technological improvement has occurred in last decade in technologies both for characterizing and treating groundwater. Coupled with the technologies themselves have been improved strategies (e.g. Triad) to decrease uncertainty and improve decision making at contaminated sites.

Much documentation available on new technologies and approaches is available on the internet largely because of the many projects being conducted using public funds in the U.S.—especially by EPA and the Departments of Energy and Defense. Nonetheless, additional cost and performance data is always needed, especially for new and emerging technologies. Continued technology advances can be expected, not only in higher technology areas like nanotechnology, but also in less intensive, more natural processes like bio and phytoremediation.

6. Additional Resources

Listed below are important internet resources devoted to the remediation of contaminated soil and groundwater:

1. http://cluin.org (or http://www.epa.gov/tio)
2. http://www.epareachit.org
3. http://www.frtr.gov
4. http://www.gwrtac.org
5. http://www.rtdf.org
6. http://www.epa.gov/ord/SITE
7. http://www.afcee.brooks.af.mil/products/techtrans/treatmenttechnologies.asp
8. http://www.itrcweb.org/
9. http://www.serdp.org/research/research.html
10. http://www.epa.gov/etv/

MASS FLUX AS A REMEDIAL PERFORMANCE METRIC AT NAPL CONTAMINATED SITES

MICHAEL D. ANNABLE
Department of Environmental Engineering Sciences,
University of Florida, Gainesville, FL 32611, USA

Abstract. Recent research efforts have shifted from a focus on contaminant concentration in soil and water, for site characterization and remedial performance assessment, to one based on contaminant mass flux and discharge. The shift has lead to advances in our ability to collect data on contaminant mass flux and discharge at field sites. The current field methods for measuring mass flux and integrated mass load include traditional transects of wells and multilevel samplers, integral pumping methods and passive flux meters. The generation of spatially discrete contaminant and Darcy flux data has increased our ability to observe flux distributions associated with complex DNAPL source zones. The characteristics of flux distributions from DNAPL contaminated sites and how this information can be used for performance assessment of remedial technologies is the focus of this paper. Changes in mass flux, as a result of remedial activities, has lead to insights on the relationship between flux architecture and source mass removal. This relationship is reviewed in light of available results from both laboratory and field experiments.

Keywords: groundwater systems, contaminant flux, performance assessment

1. Introduction

Land and water resources that have been impacted by mans activities require management strategies to protect the environment and provide long term stewardship (EPA, 2004; Abriola, 2005). Management of contaminated sites continues to advance in both site characterization methods and innovative remedial technologies. In the area of site characterization, research focus has shifted to quantifying flux of both contaminants and groundwater at field sites (Einerson and Mackay, 2001; API, 2002, 2003; ITRC, 2003; EPA 2004; NRC, 2004). As a result of this shift more approaches have been released including a method developed at the University of Florida for directly measuring water and

M.D. Annable et al. (eds.), Methods and Techniques for Cleaning-up Contaminated Sites, 177–186.

contaminant flux in wells or boreholes (Hatfield et al., 2002, 2004, Annable et al., 2005). The method uses sorbents that are inserted in a well screen to trap contaminants that move advectively with groundwater. This method, called the passive flux meter or PFM, has been developed and tested for several different contaminants (e.g., perchloroethylene, trichloroethylene, perchlorate, MTBE, etc.) at over 20 field sites including Cape Canaveral Air Station in Florida, Ft. Lewis Army Base in Washington, Hill AFB in Utah, Port Heuneme Navy Base in California, Canadian Forces Base Borden in Ontario, Edinburgh Air Base South Australia and an industrial site in Wales. Recent publications include methods for evaluating flow convergence at wells (Klammler et al., 2007) and determining flow direction (Campbell et al., 2006).

Recent advances have been made in evaluating the use of flux measurements for NAPL contaminated site assessment and management (Basu et al., 2006, Falta et al., 2005a&b). A major component of this advance has focused on establishing relationships between mass present in NAPL source areas and mass flux (or source strength) leaving the source area and forming a contaminant plume. Recent research has focused on establishing these relationships (Fure et al., 2006, Parker and Park, 2004; Falta et al., 2005, Jawitz et al., 2005; Lemke et al., 2004). This work has established the fundamental knowledge needed to advance site management strategies based on mass flux, mass discharge and mass balance. Site evaluations need to consider the age of the site, the relative mass in the source area and plume, and some estimate of plume longevity and risk given various remedial options.

Management decisions at contaminated field sites are typically based on the available data complied on subsurface conditions related to both hydrogeology and geochemistry. Final decisions are usually linked to risk drivers, regulatory requirements and future plans for site utilization. Critical to any decision on remedial strategies and long term stewardship is site characterization data quantity and quality. Site characterization data can be categorized as discrete point samples or spatially integrated quantities. Site data can also be characterized as temporally discrete or temporally averaged. Examples of these categories for water quality might include multilevel snap shot sample sets as instantaneous point samples in contrast to integral pump sample data which spatially integrates concentration. Temporally integrated samples might include diffusion based samplers or dosimeters. This paper focuses on current and emerging technologies that are designed to collect data on various spatial and temporal scales. What is the value of various types of site characterization data for decision and remedial design? The general resolution of different sampling methods for determining adequate knowledge for site management is critical to answering such questions.

In recent years, site characterization technologies have been developed to characterize contaminant mass flux and integrated mass discharge at contaminated sites. This information is deemed valuable for site management including risk assessment, quantifying natural and enhanced attenuation processes, remedial performance assessment and measuring loads to receiving bodies. The methods currently available for quantifying both local mass flux and integrated mass load can be categorized by spatial and temporal scales of measurement. Point scales techniques include multilevel samplers and passive flux meters within screened wells. Both methods collect data to determine the local mass flux [mass per unit area per time] that can be spatially integrated to determine mass discharge [mass per time]. Alternatively integral methods, such as integral pump tests, collect data averaged over fully screened wells and large volumes of water pumped from aquifers. This information, while lacking spatial resolution, has advantages in that local "hot spots" are incorporated. The utility of each scale of site flux characterization is explored in terms of site management decisions including plume or source management and remedial design improvements. The overall discussion of site characterization is linked to the value and cost of additional site characterization collected for site management purposes.

2. Contaminant Flux and Load Measurement Methods

The methods currently available for field measurement of contaminant local mass flux and integrated mass load are reviewed.

2.1. TRANSECT METHOD

The most traditional method for calculating contaminant discharge from NAPL source areas uses the product of the Darcy flux, the well cross sectional area within a transect of wells, and the contaminant concentration in the well (i.e., the "Transect Method" as described in API, 2003). The Darcy flux can be calculated as the product of independent measures of the hydraulic gradient and hydraulic conductivity. To calculate the contaminant mass flux, groundwater samples are collected from the fully screened wells forming the transect. The contaminant flux is calculated as the product of the Darcy flux for each well and the flux averaged concentration collected from each well. In this method simplification of the flux is acknowledged since average values are used for the hydraulic conductivity, hydraulic gradient and flux averaged concentration. Each of these values could have variability in space and local mass flux could be significantly different than the average values. It is assumed that the integral values used are appropriate for calculating average fluxes and integrated fluxes over the well transect. This assumption has not been verified for field sites and forms a need for future research.

The transect method can also be applied for a transect of multilevel samplers down gradient of the source area. In this case the calculation is made for each sampling location to determine the local mass flux using the local contaminant concentration. The local Darcy flux can only be calculated if the local hydraulic conductivity is measured. This may be available from locally conducted slug tests or borehole flow meter analysis in adjacent screened wells. Typically this information is not available and the site average hydraulic conductivity is used. The local hydraulic gradient, while possible to measure, is rarely determined and average values are used from monitoring wells.

In a recent study, MLS-based measurements were conducted at four sites (Guilbeault et al., 2005). In this study they found a range of integrated mass load of 41 - 85 g/day for TCE. Errors associate with transect based method are difficult to determine but in a study by Kubert and Finkel (2006) were found to be typically in the range of 20 to 50 percent.

The primary advantage of transect methods is that well established methods are used often incorporating existing infrastructure existing at the site. These methods do generate some spatial information on mass flux distributions. Estimation of uncertainty, while difficult is available and has been tested. The methods generally produce minimal waste generation. The primary disadvantages include the necessity to acquire independent measurements of the Darcy flux. This typically includes measurements for hydraulic gradient which for low gradient systems can introduce errors (Devlin and McElwee 2007). The methods interrogate a small volume of the aquifer and are instantaneous measures. Thus the data must be spatially integrated to obtain contaminant mass loads.

2.2. PASSIVE FLUX METERS

An alternate approach to the measurement of groundwater and contaminant fluxes involves deployment of a permeable, sorbent pack (i.e., PFM) in the wells located along a transect and screened across the entire saturated thickness of the aquifer (Hatfield et al., 2002, 2004; Annable et al., 2005). The PFM sorbent material is selected to capture the dissolved contaminants in the groundwater as it flows through the flux meter during the designated period of exposure under natural gradient groundwater flow conditions. Thus, flux measurements using this approach are referred to as "passive" in contrast to methods that requires pumping water.

The PFMs placed in wells are pre-saturated with "resident tracers" that are desorbed and depleted as groundwater flows through the well. Groundwater fluxes are calculated from the depletion of tracer mass, while the mass of contaminants captured on the sorbent is used to estimate contaminant fluxes. By analyzing the sorbent for the residual tracers and the captured contaminants in vertical segments of the PFM, the depth distribution of groundwater and contaminant fluxes along the well screen interval can be determined. Hatfield et al. (2004) presented the theoretical development for PFM, Annable et al. (2005) report on field-scale validation, while Basu et al. (2006) present a field-scale PFM application for site characterization (Figure 1).

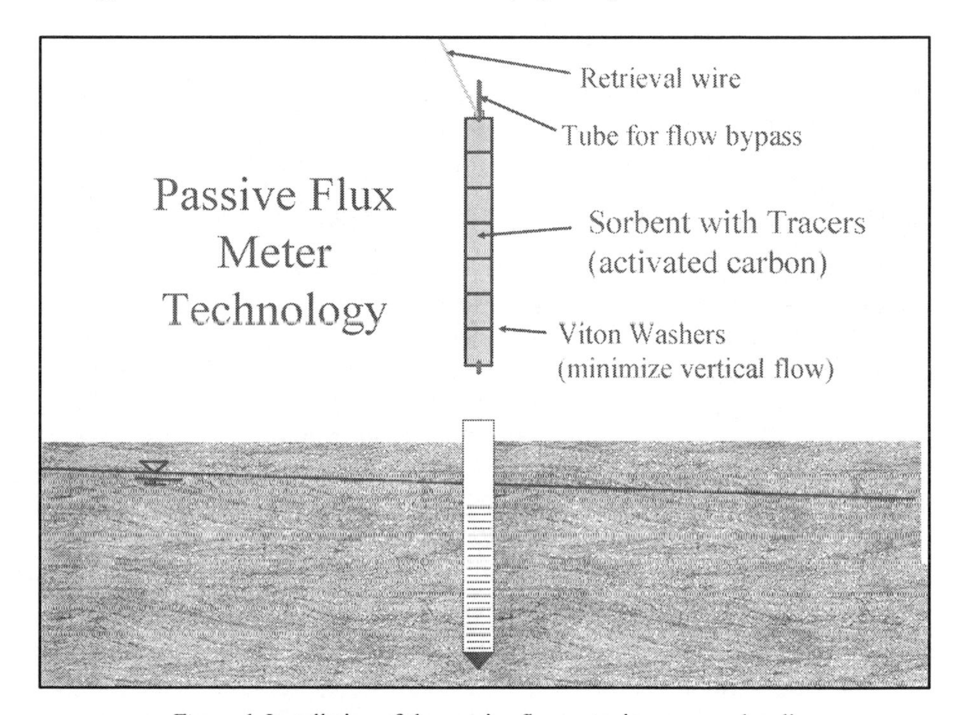

Figure 1. Installation of the passive flux meter in a screened well.

The use of PFMs can be optimized by selecting tracers with appropriate retardation on the sorbent used, and by installing the PFMs for a time period such that tracer loss and contaminant mass accumulation are quantifiable. These design parameters are based on existing estimates of the groundwater flux and contaminant concentrations at the site. More advanced interpretation of PFM data is provided by Klammler et al. (2007).

The advantages of the PFM approach is that spatial information on cumulative water and contaminant mass flux. Methods that exist to estimate uncertainty indicate that the PFM approach produces low errors in integrated mass load (Kubert and Finkel 2006). The PFM method can be used to generate local estimates of horizontal aquifer conductivity given measured hydraulic gradients. The method produces minimal solid waste volumes. The passive nature allows application in remote settings and the method does integrate information in time. The primary disadvantages include that the PFM interrogates small volumes of the aquifer thus requiring that the data be spatially integrated to obtain contaminant mass flows and total water discharge. Tracers are required to obtain Darcy flux measurements and must be approved. The systems as currently designed will not function in all wells such as rock boreholes.

To date the method has been deployed at more than 25 sites primarily government owned facilities. The largest deployment to date was at Ft. Lewis Washington where 700 linear feet was deployed in three 10 well transects. The deepest application to date has been 67 m at a landfill near Perth Australia.

2.3. INTEGRAL PUMP TESTS

The integral pump test (IPT) technique for measurement of mass flux or discharge is primarily based on measurement of the contaminant concentration-time series in the effluent of multiple pumping wells aligned perpendicular to the prevailing direction of groundwater flow. The use of concentration-time series information to estimate contaminant flux or mass discharge was initially described by Teutsch et al. (2000) and Ptak et al. (2000), and applications of the technique were described by Bockelmann et al. (2001, 2003). The measured concentration of contaminants is used with independent estimates of the natural gradient Darcy flux. In this respect the method is similar to the transect method with the difference being the significant volume of water which is interrogated during the test. This provides an integral measure of the contaminant concentration across the well transect and avoids the concern related to missing local "hot spots" within the transect.

The primary advantage of the IPT is the integral nature of the flux measurements. The method generates contaminant mass flow estimates and does supply some limited information on the spatial distribution of mass flux across the well transect. The method can be applied to deep aquifers and have no limitations based on well design. The disadvantages include the volume of waste generation is significant and can be prohibitively costly. The method can require a lengthy deployment time depending on aquifer characteristics. The method requires independent measures of the Darcy flux however these measurements can be incorporated into the pump test.

3. Remedial Performance Metrics

The underlying goal of source zone or nonaqueous phase liquid (NAPL) treatment is the reduction in risk posed by the NAPL contamination to environmental and anthropogenic resources. Aqueous contaminant concentrations can be viewed as surrogate measures of risk, and are often the primary metric for regulatory decisions. Under this traditional framework, the goal for NAPL source treatment technologies is therefore to achieve acceptable aqueous concentrations (e.g., drinking water standards) by either removing or destroying NAPL mass in the source zone. Unfortunately, current NAPL remedial technologies are unable to completely eliminate NAPL from source areas at all sites, and partial mass removal from NAPL source zones is unlikely to be sufficient to meet drinking water standards at all locations within a site. Other potential benefits may, however, be achieved as a result of source treatment, and one proposed benefit is a reduction in contaminant mass discharge (M/T), from the NAPL source area (Rao et al., 2002; EPA, 2003; Stroo et al., 2003; ITRC, 2003; NRC, 2004). The magnitude of the benefit realized needs to be quantified by measuring mass flux and discharge before and after remedial efforts are implemented.

Flux measurement should be conducted using the same protocol and differences in hydrologic conditions should be considered when interpreting the data. Also critical is the need to provide adequate time between the remedial activity and the post-remedial flux measurement to assure that conditions are representative of natural groundwater flow and contaminant dissolution from the original source area. Too often there is a rush to measure the effect of the remedial action and the measurements are only indicative of modifications still significantly influenced by remedial actions such as elevated temperatures or residual flushing agents. A general rule may be to provide at least two pore volumes of natural gradient water flow to have taken place thought the source area prior to post-remedial monitoring of mass flux. For many sites this could mean 2 to 5 years following remediation. Often site managers may find this length of time unacceptable. Partly for this reason very few studies have been completed to date with high quality measurements of pre- and post-remediation flux measurements.

4. The Link Between Source Zone Mass and Mass Flux

A number of modeling approaches have been used to evaluate whether significant reduction in mass discharge will result from source-zone mass depletion (Sale and McWhorter, 2001; Rao et al., 2002; Rao and Jawitz, 2003; Lemke et al., 2004; Parker and Park, 2004; Enfield et al., 2005; Jawitz et al.,

2005; Wood et al., 2005). The impact of reductions in mass discharge resulting from source treatment on the associated changes within the dissolved plume have also been examined in recent modeling analyses (Falta et al., 2005a,b). Results from these models suggest that a wide range of flux behavior may occur as a function of hydrogeological conditions and NAPL distributions. Limited field measurements in hydraulically isolated test cells do suggest that reductions in contaminant mass discharge are indeed observed after removal of NAPL source mass (Brooks et al., 2004). However, it may be argued that tests performed in isolated flow domains (test cells) do not represent actual field situations at NAPL sites.

5. Conclusions

These flux measurement techniques provide a method for assessing remedial performance based on mass flux and discharge from a source zone. Measurements of mass flux and discharge immediately down gradient of a source area provide an early assessment of remedial performance and an indication of the likely plume response.

Acknowledgement

The work upon which this paper is based was supported by the U.S. Environmental Protection Agency through its Office of Research and Development with funding provided by the Strategic Environmental Research and Development Program (SERDP), a collaborative effort involving the U.S. Environmental Protection Agency (EPA), the U.S. Department of Energy (DOE), and the U.S. Department of Defense (DoD). It has not been subjected to Agency review and, therefore, does not necessarily reflect the views of the Agency and no official endorsement should be inferred.

References

Abriola, L.M. 2005. Contaminant source zones: Remediation or perpetual stewardship? Environ. Health Perspectives, 113 (7): A438-A439.

Annable, M.D., K. Hatfield, J. Cho, H. Klammler, B.L. Parker, J.A. Cherry, and P.S.C. Rao. 2005. Field-Scale Evaluation of the Passive Flux Meter for Simultaneous Measurement of Groundwater and Contaminant Fluxes. Env. Sci. Tech., 39 (18): 7194-7201.

API 2003. "Groundwater Remediation Strategies Tool." Regulatory Analysis and Scientific Affairs Department, Publication No. 4730, American Petroleum Institute, Washington DC.

API 2002. "Estimating Mass Flux for Decision-Making: An Expert Workshop." American Petroleum Institute, Washington DC.

Basu, N.B., P.S.C. Rao, I.C. Poyer, M.D. Annable, and K. Hatfield. 2006. Flux-Based Assessment at a Manufacturing Site Contaminated with Trichloroethylene, Journal of Contaminant Hydrology, Vol. 86, No. 1-2, pp. 105-127.

Bockelmann, A., T. Ptak, and G. Teutsch. 2001. An analytical quantification of mass fluxes and natural attenuation rate constants at a former gasworks site. J. Contam. Hydrol., 53, 429-453.

Bockelmann, A., D. Zamfirescu, T. Ptak, P. Grathwohl, and G. Teutsch. 2003. Quantification of mass fluxes and natural attenuation rates at an industrial site with a limited monitoring network: a case study. J. Contam. Hydrol., 60 (1-2): 97-121.

Brooks, M.C., M.D. Annable, P.S.C. Rao, K. Hatfield, J.W. Jawitz, W.R. Wise, A.L. Wood., and C.G. Enfield. Controlled Release, Blind Test of DNAPL Remediation by Ethanol Flushing. Journal of Contaminant Hydrology, Vol. 69, 2004, pp. 281-297.

Campbell, T.J., K. Hatfield, H. Klammler, M.D. Annable, and P.S.C. Rao. Magnitude and Directional Measures of Water and Cr(VI) Fluxes by Passive Flux Meter, Env. Sci. Tech.. Vol. 40, 2006, pp. 6392-6397.

Devlin, J.F. and C.D. McElwee. 2007. Effects of measurement error on horizontal hydraulic gradient estimates. Ground Water, 45 (1): 62-73.

Einarson, M.D. and D.M. Mackay. 2001. Predicting impacts of groundwater contamination. Environmental Science and Technology. 35 (3): 66A-73A.

Enfield, C.G., A.L. Wood, M.C. Brooks, M.D. Annable, and P.S.C. Rao. Design of aquifer remediation extraction systems: (1) Describing hydraulic structure and NAPL architecture using tracers. Journal of Contaminant Hydrology, 81 (1-4): 2005, pp. 125-147.

EPA 2004. The DNAPL Remediation Challenge: Is There a Case for Source Depletion? Kavanaugh, M.C. and P.S.C. Rao (editors), EPA/600/R-03/143, National Risk Management Research Laboratory, Office of Research and Development, U.S. Environmental Protection Agency, Cincinnati, Ohio, USA.

Falta, R.W., P.S.C. Rao, and N. Basu. 2005a. "Assessing the Impacts of Partial Mass Depletion in DNAPL Source Zones: I. Analytical Modeling of Source Strength Functions and Plume Response." Journal of Contaminant Hydrology, 78 (4): 259-280.

Falta, R.W., N. Basu, and P.S.C. Rao. 2005b. "Assessing Impacts of Partial Mass Depletion in DNAPL Source Zones: II. Coupling Source Strength Functions to Plume Evolution," Journal of Contaminant Hydrology, 79 (1-2): 45-66.

Fure, A.D., J.W. Jawitz, and M.D. Annable. 2006. DNAPL Source Depletion: Linking Architecture and Flux Response, Journal of Contaminant Hydrology, 85: 118-140.

Guilbeault, M.A., B.L. Parker, and J.A. Cherry. 2005. Mass and flux distributions from DNAPL zones in sandy aquifers. Ground Water, 43 (1): 70-86.

Hatfield, K., M.D. Annable, J. Cho, P.S.C. Rao, and H. Klammler. 2004. "A Direct Passive Method for Measuring Water and Contaminant Fluxes in Porous Media." Journal of Contaminant Hydrology, 75 (3-4), 155-181.

Hatfield, K., P.S.C. Rao, M.D. Annable, and T. Campbell. 2002. Device and method for measureing fluid and solute fluxes in flow systems, Patent US 6,402,547 B1, US Patent Office, Washington DC.

ITRC 2003. "Assessing the performance of DNAPL source reduction remedies." Dense nonaqueous phase liquids team, Interstate technology & regulatory council.

Jawitz, J.W., A.D. Fure, G.G. Demmy, S. Berglund, and P.S.C. Rao. 2005. Groundwater contaminant flux reduction resulting from nonaqueous phase liquid mass reduction. Water Resour. Res. 41 (10): Art. No. W10408.

Klammler, H., K. Hatfield, and M.D. Annable. Concepts for Measuring Horizontal Groundwater Flow Directions Using the Passive Flux Meter, Advances in Water Resources, 30 (4): 984-997, 2007.

Klammler, H., K. Hatfield, M.D. Annable, E. Agyei, B.L. Parker, J.A. Cherry, and P.S.C. Rao. 2007. General analytical treatment of the flow field relevant to the interpretation of passive fluxmeter measurements, Water Resour. Res., 43, W04407, doi:10.1029/2005WR004718.

Kübert, M. and M. Finkel. 2006. Contaminant Mass Discharge Estimation in Groundwater Based on Multi-level Point Measurements: A Numerical Evaluation of Expected Errors, J. Contam. Hydrol., 84 (1-2), 55-80.

Lemke L.D., L.M. Abriola, and J.R. Lang. 2004. Influence of hydraulic property correlation on predicted dense nonaqueous phase liquid source zone architecture, mass recovery and contaminant flux. Water Resour. Res. 40 (12): Art. No. W12417.

NRC 2004. Contaminants in the Subsurface: Source Zone Assessment and Remediation. The National Academies Press, Washington D.C.

Parker, J.C. and E. Park. 2004. Modeling field-scale dense nonaqueous phase liquid dissolution kinetics in heterogeneous aquifers. Water Resour. Res. 40, W05109, doi; 10.1029/ 2003WR002807.

Patterson, B.M., G.B. Davis, and C.D. Johnston. 1999. Automated in situ devices for monitoring of VOCs and oxygen in water and soil environments. In: Contaminated Site Remediation: Challenges posed by urban and industrial contaminants (Ed. C.D. Johnson). Proc. Contaminated Site Remediation Conference, Fremantle, Western Australia, pp. 227-234.

Patterson, B.M., G.B. Davis, and A.J. McKinley. 2000. Volatile organic compounds in groundwater, probes for the analysis of, In Encyclopedia of Analytical Chemistry, (Ed. R.A. Meyers), John Wiley & Sons Ltd, Chinchester, pp. 3515-3526.

Ptak, T., M. Schirmer, and D. Teutsch. 2000. Development and performance of a new multilevel groundwater sampling system. Risk, Regulatory and Monitoring Considerations: Remediation of Chlorinated and Recalcitrant Compounds, Wickramanayake, Godage B., Gavaskar, Arun R., Kelley, Mark E. and Nehring, Karl W. (eds.), Battelle Press, Columbus, Ohio, USA, ISBN 1-57477-095-0, 95-102.

Rao, P.S.C., J. Jawitz, C. Enfield, R. Falta, M. Annable, and A. Wood. 2002. Technology integration for contaminated site remediation: Cleanup goals and performance criteria. In Groundwater Quality 2001. Proceedings, Edited by S. Thornton and S. Oswald, IAHS Publication no. 275, pp. 571-578.

Rao, P.S.C. and J.W. Jawitz. 2003. Comment on “Steady state mass transfer from single-component dense nonaqueous phase liquids in uniform flow fields” by T.C. Sale and D.B. McWhorter. Water Resour. Res., 39, 1068.

Sale, T.C. and D.B. McWhorter. 2001. Steady-state mass transfer from single-component dense non-aqueous phase liquids in uniform flow fields, Water Resour. Res., 37, 393-404.

Stroo, H.F., M. Unger., C.H. Ward., M.C. Kavanaugh., C. Vogel, A. Leeson, J.A. Marqusee and B.P. Smith. 2003. Remediating chlorinated solvent source zones. Environmental Science & Technology, 37 (11): 224A-230A.

Teutsch, G., T. Ptak, R. Schwarz, T. Holder. 2000. Ein neus integrales Verfahren zur Quantifizierung der Grundwasserinnission: I. Theoretische Grundlagen. Grudwasser 4 (5): 170-175.

Wood, A.L., C.G. Enfield, M.D. Annable, M.C. Brooks, P.S.C. Rao, D. Sabatini, and R. Knox. 2005. Design of aquifer remediation extraction systems: (2) Estimating site-specific performance and benefits of partial source removal. Journal of Contaminant Hydrology, 81 (1-4), pp. 148-166.

CARBON/POLYMER COMPOSITE ADSORPTION-FILTERING MATERIALS FOR INDIVIDUAL PROTECTIVE SYSTEMS

MYKOLA T. KARTEL*,
Institute of Sorption and Problems of Endoecology, NAS of Ukraine;
YURIY V. SAVELYEV,
Institute of Macromolecular Chemistry, NAS of Ukraine;
NICKOS KANELLOPOULOS,
Institute of Physical Chemistry, NCSR "Demokritos", Greece.

Abstract. The aim of this work was the development of improved adsorption-filtering materials appropriate for individual or collective protection systems. The work focused on creating and studying novel composite adsorption-filtering materials (CAFM) based on 'Carbon Adsorbent [active component] - Polymer [matrix]' and optimizing the media characteristics. The following different methods for preparing optimized CAFMs were pursued: a) formation of block CAFM (hard structure); b) formation of an elastic CAFM based on polymer foams; c) and formation of an immobilized CAFM (adsorption-active components were immobilized on a polymeric filter material by the extrusion process). The optimized composites can be used as a means of protection for personnel, working in areas polluted by toxic (including radioactive) gases and dusts.

Keywords: composite; adsorption; filtering; protection

1. Introduction

A target of this work was to develop novel adsorption-filtering materials (CAFM) and to optimize their properties. Means of creating composite materials on the basis of carbon adsorbents and polymeric matrixes capable of carrying out simultaneous adsorption and filtering functions at the point of use in gas and

* To whom correspondence should be addressed. Mykola Kartel, Institute of Sorption and Problems of Endoecology, NAS of Ukraine, General Naumov street 13, 03164, Kiev, Ukraine; e-mail: kartel@mail.kar.net

M.D. Annable et al. (eds.), Methods and Techniques for Cleaning-up Contaminated Sites, 187–196.

liquid environments were developed (Marushko 1979). This research was conducted in this field developing appropriate technologies and preparation of new prospective adsorption-filtering materials in the framework of international cooperation (NATO grant SfP 977995).

An adsorption-active component the fruit-stones active carbon of KAU-type and its modifications were considered.

The polymeric systems include various commercial polymers – polyvinyl alcohol (PVA), polyvinyl chloride (PVC), linear polyurethane's (PU) and fiber polypropylene (PP), which formed the basis for composites of block, foamy and elastic characteristics. Appropriate modification of polymeric matrixes as carrier materials can also have additional functions, for example, bactericide or fungicide properties, complexion ability to metal ions, etc.

Physical and chemical properties of composites (adsorption activity, volume and distribution of pores volume, specific surface area) as well as their operational characteristics (gas and aerosol permeability, flow resistance, etc.) were investigated depending on ratio of carbon to polymer. The variants of use of such composites were considered as a basis to produce the improved filters and respirators.

2. Experimental

Three types of CAFMs have been generated - block and foamy types, and also immobilized particles of adsorbent on the non-cloth filtering material.

As a base carbon adsorbent, fruit-stone activated carbon KAU has been used. The generation techniques were developed at the Institute of Sorption and Problems of Endoecology, NAS of Ukraine (National Standard TU U 88.290.015-94). The fractions of carbon used were 0.5-1 mm (grains) and less than 0.1 mm (powder) in diameter. The main adsorption characteristics of the material: volume of adsorption pores of *benzene* (W_s) 0.6 cm^3/g; specific surface area on *argon* (S_{sp}) 1100 m^2/g.

As polymeric components for generation of composites of block type, polymers of industrial production PVA, PVC and PU were used. To prepare a CAFM of a foamy type, foamed polyurethane (FPU) was used. Immobilized composites were generated based on industrial non-cloth filtering material (NFM) from PP ultrafine fibers.

Method of generation of block CAFM. Active carbon KAU was initially processed using solvent (carbon tetrachloride, ethylbuthylacetate, methylene chloride, etc.), followed by drying at room temperature. After this, carbon was mixed with solutions of polymers of different concentration (5-30% wt.), placed in special "sieve" form and dried under loading (0.1-1 kg/cm^2) at temperature 50-60°C and the lowered pressure (2 mm Hg). The resulting ratio of carbon to polymer in the block CAFM was 85-99.8 : 0.2-15, % wt. (Patent 2002).

Method of generation of elastic (foamy) CAFM. Active carbon KAU was initially processed using solvent, dried at room temperature and mixed with the components necessary for producing a foamed PU. After this, formation and drying of the material was carried out. The resulting ratio of carbon to polymer in the elastic CAFM was 25-55 : 45-75, % wt. (Patent 2003a).

Method of generation of immobilized CAFM. The powder of high disperse active carbon KAU or its suspension in a polymer (PVA, PU) was delivered through a dosing device into a system of aerodynamic formation of ultrafine fibers from the melted PP simultaneously drawing components on a moving generation surface, where these were coupled. The resulting ratio of carbon to polymer in the immobilized CAFM was 15-85 : 85-15, % wt. (Patent 2003b).

Parameters related to the porous structure of initial carbon and the resulting CAFMs have been investigated including the parameters W_s, and S_{sp}. The distribution of volumes of macro- and mesopores on equivalent radii were determined with a mercury porosimeter (Model M9200, "Cultronics").

Research on kinetic and diffusion characteristics of carbon, KAU and CAFMs were carried out under static conditions and in a dynamic mode using a substance - marker (*methylene blue*) from corresponding water solutions. Also, the significance of the time to achieve adsorption for effective and true diffusion were determined and their comparative analysis was completed.

Operational properties of CAFM were estimated by the level of dust content (dust creation) for active carbon and composites, and on dynamic characteristics (time of protective action on *benzene* in accordance with GOST 12.4.158-75, resistance to a constant flow of *air* in accordance with GOST 10188-74, factors of penetration of *microgrinding powder M-5* and an *oil fog* in accordance with GOST 12.4.156-75).

3. Results and Discussion

By original technologies three principally different CAFMs derived from fruit stones carbon KAU (grains – KAU$^{(1)}$ and powder – KAU$^{(2)}$) and polymers of various structure (Table 1) were synthesized and investigated, namely: block (KAU-PVA, KAU-PVC, KAU-PU), elastic foamy (KAU-FPU) and immobilized (KAU-NFM) CAFMs.

A block CAFM was prepared from carbon grains suspended in solutions of polymers. On Figure 1 the relationship between polymer content in composite (C) versus initial concentration of polymer in solution (C*) and loading at formation of composite (P) are presented.

The synthesis of elastic foamy CAFM involved several chemical transformations in carbon grains or powder suspension with solution of urea, diisocyanate, chain longer and polyester:

- reaction of chain's growth

  2OCN-R-NCO + HO-R'-OH → OCN-R-NHCOO-R'-OOCNH-R-NCO

  2OCN-R-NCO + H-R'-H → OCN-R-NHCO-R'-OCNH-R-NCO

- reaction of gas-educing

  R-NHCOOH → $R\text{-}NH_2 + CO_2$

- formation of the bonds with cross-links

  -HN-CO-NH- + OCN-R-NCO + -HN-CO-NH- →

  -HN-CON*-CONH-R-NHCO-N*CO-NH-

where: R – various radicals, N* - free valences were able to bond with carbon.

The immobilization of carbon particles on the non-fabric filtering material was produced during the formation of microfibers from melted PP using airflow technology. A principal scheme of producing CAFM of this type is presented in Figure 2.

By adsorption methods, weight analysis and *mercury* porosimetry, change in porous structure of CAFMs were investigated. It was determined that polymers acted basically as a blocking system to transport by pores in the carbon (macro- and partially mesopores). Intervals of insignificant influence of polymers on adsorption parameters of active carbon in CAFM can be determined from the data presented in Figures 3, 4 and Table 2.

TABLE 1. Characterization of polymeric components to prepare CAFM with carbon KAU.

Polymer matrix	Elementary chain	Solvent
PVA	$(\text{-}CH_2\text{-}CHOH\text{-})_n$	H_2O
PVC	$(\text{-}CH_2\text{-}CHCl\text{-})_n$	DMFA
PU (PU-1, PU-2, PU-3)	$(R_1\text{-}CONH\text{-}R_2\text{-}NHCO\text{-}R_3\text{-}OCONH\text{-}R_2\text{-}NHCO\text{-}R_1\text{-})_n$, where $R_{1\text{-}3}$* – polyesters, diisocyanates, chain longer	DMFA, H_2O
PP	$(\text{-}CH_2\text{-}CHCH_3\text{-})_n$	-

where R_1*:

PU-1: $[(CH_3)_2N^{+}(NH_2)CH_2CH(OH)CH_2OH]Cl^{-}$

PU-2: $(CH_3)_2N\text{-}NH_2$

PU-3: $[(CH_3)_2N^{+}(NH_2)CH_2CONHNH_2]Cl^{-}$

R_2*: $\text{-}C_6H_4\text{-}CH_2\text{-}C_6H_4\text{-}$

R_3*: $\text{-}[\text{-}(CH_2)_4\text{-}O\text{-}]_n\text{-}$

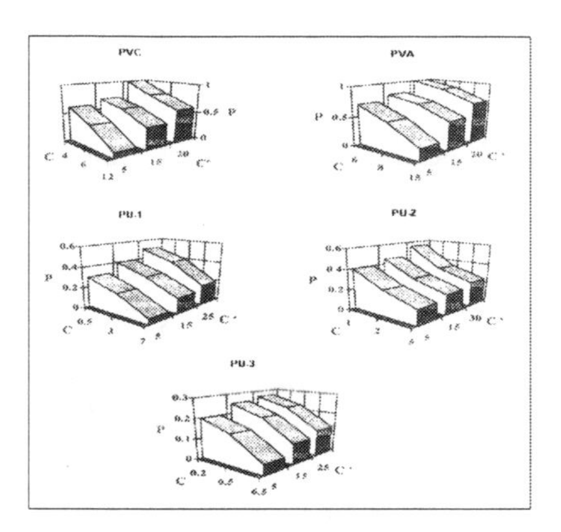

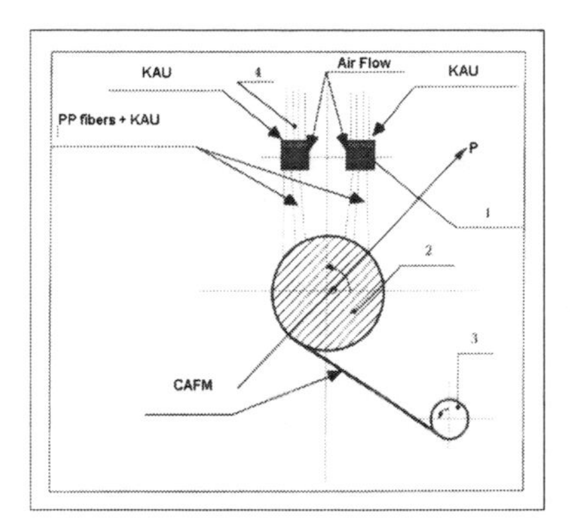

Figure 1. Diagrams of synthesis of block CAFM: C – content of polymer in composite, % wt.; C* - initial concentration of polymer in solution, % wt.; P – loading at formation of composite, kg/cm^2.

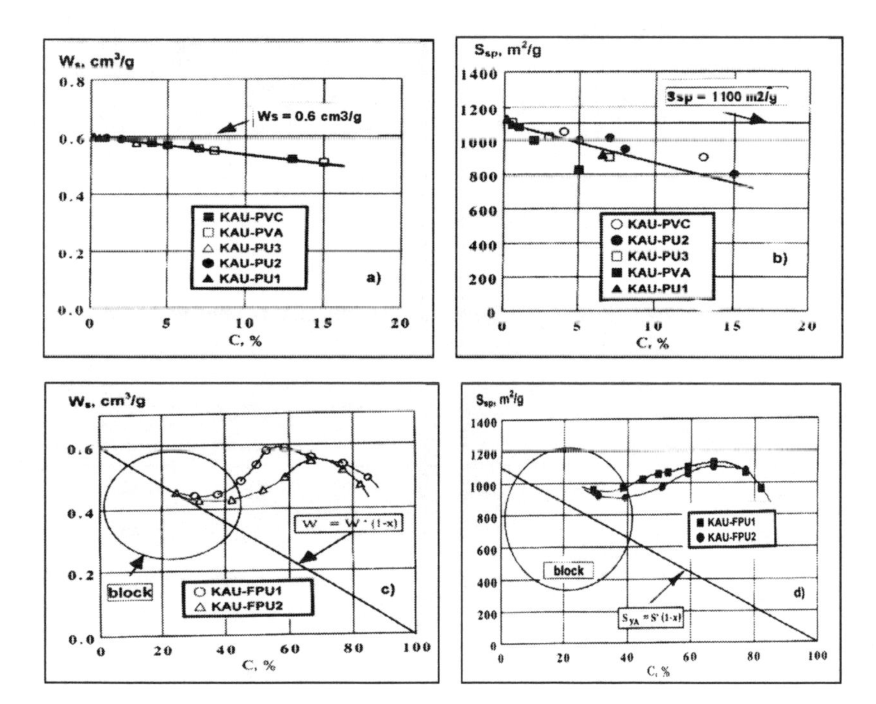

Figure 2. A principal scheme of producing CAFM of immobilized type on the base of carbon KAU (powder) and PP ultrafine fibers.

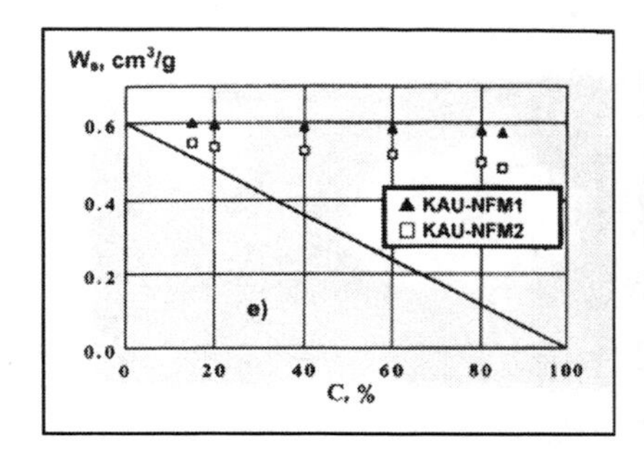

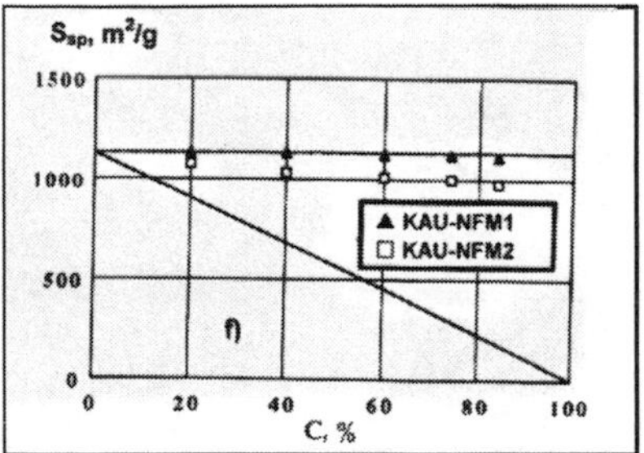

Figure 3. Significances of W_s (a, c, e) and S_{sp} (b, d, f) for carbon KAU in CAFM of various types: block (a, b), elastic foamy (c, d), and immobilized samples (e, f).

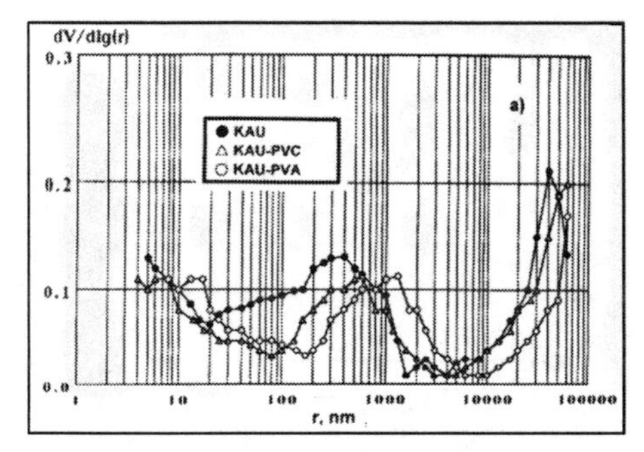

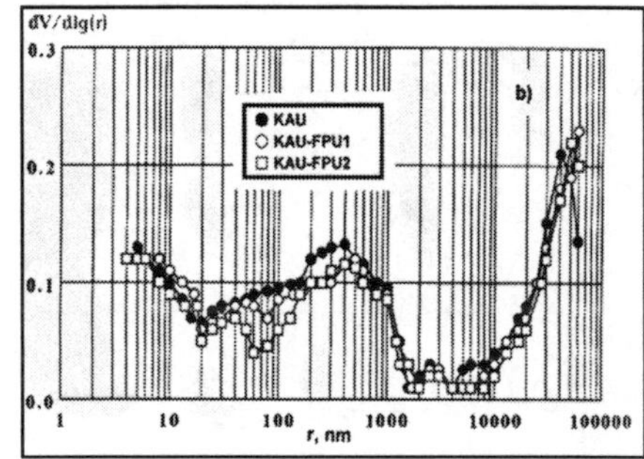

Figure 4. Distributions of pore volume on equivalent radii from mercury porosimeter study of carbon KAU and CAFM of block (a) and foamy (b) types.

TABLE 2. Changing parameters of porous structure of carbon KAU in block CAFM (1.5% wt. of polymer) and foamy CAFM (60% wt. of polymer).

Parameter	KAU(1)	Rigid blocks		Elastic foams	
		KAU(1)-PU-1	KAU(1)-PU-3	KAU(1)-FPU-1	KAU(2)-FPU-1
Volume of sorption pores (W_s), cm^3/g	0.60	0.59	0.58	0.59	0.58
Total volume of pores, cm^3/g	0.68	0.64	0.61	0.64	0.62
Volume of macropores, cm^3/g	0.22	0.19	0.18	0.19	0.17
Volume of mesopores, cm^3/g	0.24	0.23	0.21	0.23	0.23
Volume of micropores, cm^3/g	0.22	0.22	0.22	0.22	0.22
Predominant radius of pores, nm	0.72	0.88	0.91	0.88	0.90
Specific surface area (S_{sp}), m^2/g	1100	1050	1000	1080	1050
Surface area of transport pores, m^2/g	16.7	10.4	9.8	12.7	12.4

Based on the data obtained, the following conclusions result:

- In block composites, a polymer content up to 10% wt. linearly reduced W_s and S_{sp}, and reduced parameters proportionally to increased contents of the polymer in the CAFM;
- In foamy composites a polymer content within 50-70% wt. reduced parameters W_s and S_{sp} of the carbon in the composite near 3-5% wt.;
- Composites of immobilized type were characterized by almost full absence of influence of polymer content on adsorption characteristics of carbon in a composite (reduction of W_s and S_{sp} at any carbon to polymer ratio within the limits of 2-5% wt.).

The kinetics of adsorption was investigated using a model substance, *methylene blue,* from water solutions on CAFMs of block and foamy types and it was determined that the diffusion characteristics of these materials and found a delayed adsorption character on the composites comparable to the initial carbon. The corresponding decrease in diffusion coefficients were on average 5-20% wt. (Table 3).

TABLE 3. Effective and factual coefficients of diffusion at adsorption of *methylene blue* from aqueous solutions by carbon KAU and block and foamy CAFM on its base.

Sample	D_e*10^{-8}, cm^2/s (percentage from initial)	D_f*10^{-6}, cm^2/s (percentage from initial)
KAU(1)	3.10 (100%)	5.80 (100%)
KAU(1)-PVC	2.50 (81%)	5.20 (89%)
KAU(1)-PVA	2.60 (84%)	5.22 (90%)
KAU(1)-PU-1	2.70 (87%)	5.40 (93%)
KAU(1)-PU-2	2.60 (84%)	5.30 (91%)
KAU(1)-PU-3	2.70 (87%)	5.40 (93%)
KAU(1)-FPU	2.64 (85%)	5.51 (95%)
KAU(2)-FPU	2.48 (80%)	5.39 (93%)

Standard methods were used to estimate the operational characteristics of the resulting CAFMs. It was shown that these materials differed at high level of absorption of VOCs showing better physical mechanics characteristics than the initial carbon adsorbent, thus an increased durability, a low level of dust emission (in foamy - completely was absent) and insignificant resistance to a gas flow (Table 4).

The CAFM of the immobilized type was determined to be the simplest adsorption-filtering respirator for breathing protection of people simultaneously from VOC vapors, toxic or poisonous gases, dust particles and aerosols of various structure, including radioactive (see Table 5). Such a respirator can be considered an inexpensive mini-gas mask for short-term use designed for mass

TABLE 4. Exploitation characteristics of carbon KAU and block and foamy CAFM on its base.

Sample	Contents of polymer, %	Resistance to air flow, mm H_2O	Durability at pressing, kg/cm^2	Time of protective action on *benzene*, min	Dust particles emission, %
$KAU^{(1)}$	0	10.5	-	55	15
$KAU^{(1)}$-PVC	13	12	5	40	< 0.1
	5	11	5	45	0.1
	4	10.5	4.5	45	0.15
$KAU^{(1)}$-PVA	15	11	5	45	< 0.1
	7	10.5	5	45	0.1
	6	10	4.5	45	0.1
$KAU^{(1)}$-PU-1	7	10	4	45	0.1
	3	9	3.5	50	0.1
	0.5	9	3	50	0.1
$KAU^{(1)}$-FPU	50	0.65	-	30	Absence
$KAU^{(2)}$-FPU	50	0.55	-	25	Absence

TABLE 5. Comparison of exploitation characteristics of respirators prepared on the base of commercial (non-cloth PP fibers) filtering material NFM and developed immobilized CAFM (KAU-NFM).

Parameter	Commercial respirator on a base of NFM	Experimental respirator on a base of KAU-NFM
Coefficient of penetration of microgrinding powder, %		
D = 0.28-0.34 μ	1.4	1.4
D = 2 μ	0.8	0.8
Standard aerosol (M-5)	0.5	0.04
Resistance to air flow, mm H_2O		
at V = 30 L/min	< 2	< 2
at V = 3 L/min	0.60	0.48
Time of protective action on benzene		
at C = 50 mg/m^3, h	0	8 ± 1
at C = 10 g/m^3, min	0	5 ± 1
Protective action on radioactive ^{131}I -vapors, %	< 5	85 ± 5

use in extreme situations: collapses, fires, acts of terrorism, etc. (Figure 5). Further improvement of the characteristics of adsorption-filtering materials such as «active carbon - polymer» is possible by improving the carbon adsorbent (providing ion exchange, complexion, catalytic, antibacterial and other properties), and by improving a polymeric part (improving aero- and hydrodynamic characteristics, giving complexion, bactericidal or fungicidal properties, etc.).

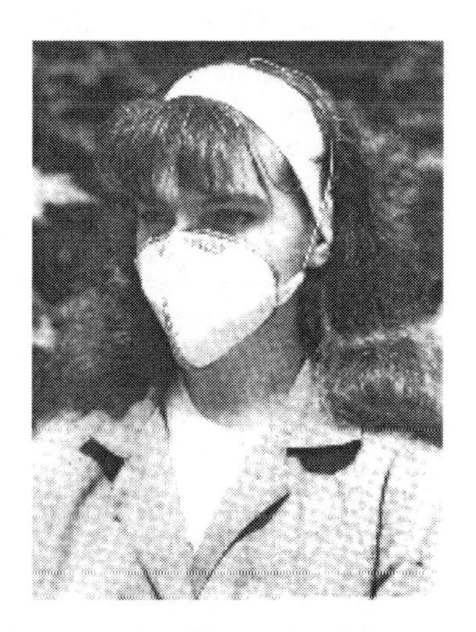

Figure 5. Simplest respirators and filters produced on the base of CAFM of immobilized type (carbon KAU/PP ultrafine fibers non-cloth material).

4. Conclusions

1. The new techniques for preparation of carbon - polymeric adsorption-filtering materials of three types - block, foamy and immobilized ones have been developed and patented.

2. The character of porous structure changes for active carbon KAU in CAFMs of all types was investigated; the mixtures of carbon and polymer were determined that were expedient to use the resulting composites as adsorption-filtering materials.
3. The kinetic characteristics of adsorption on the resulting composites were investigated. It was shown that the coefficients of diffusion for carbon KAU were reduced in composites no more than 20%.
4. The estimation of the operational characteristics has shown that prepared samples of CAFM have a lot of the improved parameters, in comparison with used (commercial) analogues such as filtering materials.
5. The results for the CAFM of the immobilized type showed the simplest means of individual protection (filters and respirators), in particular adsorption-filtering respirators like the mini-gas masks as a quite inexpensive means for short-term protection of the population in variant of mass use at some extreme situations accompanied by pollution of air by toxic and radioactive gases and aerosols.

Acknowledgement

This work was supported by NATO - grant SfP 977995 “Novel Adsorption-Filtering Materials for Individual Protective Systems”.

References

Marushko, S.Z., *Synthesis and Properties of Composite Adsorption Materials "Active Carbon/Polymer"* (Thesis Manuscript, T. Shevchenko Kiev National University, Kiev, 1999).
Patent of Ukraine N 25473. Carbon-polymeric composite adsorption material and method of its obtaining. S.Z. Marushko, M.T. Kartel, Yu.V. Savelyev, S.V. Kuznetsov, bul. N4 (2002).
Patent of Ukraine N 29034. Carbon-polymeric elastic composite adsorption material and method of its obtaining. S.Z. Marushko, M.T. Kartel, Yu.V. Savelyev, L.A. Markovskaya, bul. N4 (2003).
Patents of Ukraine NN 54404, 54405. Carbon-polymeric composite adsorption material and method of its obtaining. S.Z. Marushko, M.T. Kartel, M.V. Polyakov, V.V. Strelko et al., bul. N3 (2003).